# 离散数学

（英文版）

# Discrete Mathematics

编 著：刘红美

编 者：刘红美　王建芳　张钦

华中师范大学出版社

## 内容简介

本书是信息与计算科学和计算机科学核心课程——离散数学的基础教材。全书共分七章，分别介绍了离散数学的最基本内容：命题逻辑、谓词逻辑、集合理论、关系、图论、树和代数结构。内容叙述严谨，推理详尽。

本书可作为普通高等学校信息与计算科学专业和计算机专业学生离散数学课程的双语教学教材，亦可作为自动控制、电子工程、管理科学等有关专业的教学用书和工作人员的阅读参考。

## 新出图证(鄂)字 10 号

### 图书在版编目(CIP)数据

离散数学＝Discrete Mathematics：英文/刘红美 编著.
—武汉：华中师范大学出版社,2013.1(2020.9 重印)
ISBN 978-7-5622-5901-5

Ⅰ.①离… Ⅱ.①刘… Ⅲ.①离散数学—高等学校—教材—英文 Ⅳ.①O158

中国版本图书馆 CIP 数据核字(2013)第 002839 号

### 离 散 数 学
（英文版）
### Discrete Mathematics

作者：刘红美 ©

| | | |
|---|---|---|
| 责任编辑：袁正科 | 责任校对：罗　艺 | 封面设计：胡　灿 |
| 编辑室：第二编辑室 | 电话：027-67867364 | |

出版发行：华中师范大学出版社
社址：湖北省武汉市珞喻路 152 号
电话：027-67861549
传真：027-67863291
网址：http://press.ccnu.edu.cn　　　电子信箱：press@mail.ccnu.edu.cn
印刷：武汉邮科印务有限公司　　　督印：刘　敏
字数：318 千字
开本：787mm×960mm　1/16　　　印张：16.25
版/印次：2020 年 9 月第 1 版第 3 次印刷
定价：42.00 元

欢迎上网查询、购书

敬告读者：欢迎举报盗版，请打举报电话 027-67867353

# 前　言

为了适应经济全球化和应对科技革命的挑战，全国很多高校相继开设外语或双语教学的专业课程。离散数学在内容、学时、年级等方面有着特殊性，特别适合改革试点课程进行外语或双语教学。

离散数学是研究离散变量的结构及相互关系的一门学科，是计算机科学与技术专业的核心课程，也是其他相关专业的必修课程或重要选修课程。离散数学强调逻辑性和抽象性，注重概念、方法和应用。它的基本概念、基本理论和基本方法大量地应用在数字电路、算法分析与设计、数据结构、程序设计原理、操作系统、数据库系统、人工智能、计算机网络等专业课程中。通过对离散数学的学习，一方面为学生的专业课学习、软件开发和科学研究打下坚实的基础；另一方面培养和提高了学生的抽象思维、逻辑思维和归纳构造能力；同时也可极大地提升学生的数学建模能力，极大地提升学生进行应用研究和解决实际问题的能力。用英语或双语进行离散数学的教学，能够使学生在获得上述专业技能的同时，在专业英语方面也得到锻炼，对学生以后的职业生涯有莫大的好处。

本书作为一个学期离散数学英语教学的教材，是作者经过多年教学实践，对讲义反复增删提炼的结晶。考虑到课时的限制，本书仅包含离散数学的最基本内容：命题逻辑、谓词逻辑、集合理论、关系、图论、树和基本的代数结构。

本书力求语言流畅、通俗易懂、简明精练。为方便初次接触数学英语的同学，书中配有部分常用专业词汇的中文解释。本书适合作为普通高等学校计算机类和工程类本、专科生的离散数学外语或双语教学教材，亦可作为其他相关专业的教学用书和工作人员的阅读参考用书。

在编写过程中，作者参阅了大量国内外目前广泛使用的优秀原版外文教材和优秀中文教材以及其他相关文献和资料。书中一些定理的证明，以及部分例题和练习题采用自 Kenneth H. Rosen 的"Discrete Mathematics and Its Application (Fifth Edition)"，James L. Hein 的"Discrete Mathematics"，J. A. Bondy 和 U. S. R. Murty 的"Graph Theory with Applications"，Kenneth H. Rosen 的

"Handbook of Discrete and Combinatorial Mathematics",耿素云和屈婉玲编著的《离散数学》等著作,同时作者根据国内双语教学的实际做了恰当的取舍和编排。另外,河南理工大学计算机科学与技术学院的王建芳老师、三峡大学理学院的熊新华、张钦老师对本书部分章节的修改做了大量的工作,在此向他们表示衷心的感谢。在本书的出版过程中,得到了评审专家的精心指导和编辑的大力支持,借此机会也向他们表示衷心的感谢。

限于作者的学识水平,错误和不当在所难免,诚挚欢迎批评指正。

作 者
2012 年 9 月

# CONTENTS

1. **Propositional Logic** ································ (1)
   1.1 Propositions and Connectives ················ (1)
   1.2 Propositional WFF and Assignment ············ (10)
   1.3 Propositional Equivalences ·················· (15)
   1.4 Disjunctive Normal Form ····················· (23)
   1.5 Functionally Complete Set of Logical Connectives ················ (32)
   1.6 Rules of Inference ·························· (34)

2. **Predicate Logic** ···································· (43)
   2.1 Predicates and Quantifiers ·················· (43)
   2.2 Well-Formed Formulas in Predicate Logic ····· (53)
   2.3 Equivalent Formulas ························· (60)
   2.4 Prenex Normal Form ·························· (67)
   2.5 Inference Rules in Predicate Calculus ······· (71)

3. **Set Theory** ······································· (78)
   3.1 Sets ········································· (78)
   3.2 Set Operations ······························ (83)
   3.3 Inclusion-Exclusion ························· (90)

4. **Relations** ········································ (96)
   4.1 Cartesian Products and Relations ············ (96)
   4.2 Properties of Relations ····················· (101)

4.3　Representing Relations ………………………………………… (107)
　4.4　Closure of Relations …………………………………………… (115)
　4.5　Equivalence Relations ………………………………………… (126)
　4.6　Partial Orderings ……………………………………………… (134)

## 5. Graphs ……………………………………………………………… (149)
　5.1　Graph Terminology …………………………………………… (149)
　5.2　Representing Graphs and Graph Isomorphism ……………… (157)
　5.3　Subgraphs ……………………………………………………… (165)
　5.4　Euler and Hamilton Paths …………………………………… (174)
　5.5　The Shortest-Path Problem ………………………………… (189)
　5.6　Planar Graphs ………………………………………………… (193)

## 6. Trees ………………………………………………………………… (200)
　6.1　Basic Concepts ………………………………………………… (200)
　6.2　Roots and Orderings ………………………………………… (203)
　6.3　Spanning Trees ………………………………………………… (212)

## 7. Algebra Structures ……………………………………………… (222)
　7.1　Basic Concepts ………………………………………………… (222)
　7.2　Groups …………………………………………………………… (226)
　7.3　Rings and Fields ……………………………………………… (231)
　7.4　Boolean Algebras ……………………………………………… (237)

## Reference ……………………………………………………………… (253)

# 1
# Propositional Logic

Logic is the basis for distinguishing what may be correctly inferred from a given collection of facts. Propositional logic, which is also called propositional calculus, studies logical propositions and their combinations using logical connectives. Propositional logic is also named as zero-order logic because it does not use quantifiers, namely quantifiers range over nothing. This chapter defines the meaning of the symbolism and gives various propositional logical properties that are usually used without explicit mention. Only two-valued logic is studied in this chapter, i. e., each statement is either true or false. Multi-valued logic, in which statements have one of more than two values, involves fuzzy sets theory whose detailed discussion is out of the scope of the book. Propositional logic has vast area of applications ranging from natural science to social science. Later in this chapter, some examples of applications in computer science such as circuit design and verification of computer program correctness are presented.

## 1.1 Propositions and Connectives

In mathematical reasoning we need to use declarative sentences to state conditions and conclusions. These sentences are called propositions, which are the basic building blocks of logic. A **proposition** is a declarative sentence that is either true or false, but not both. We use the lower case letters of the alphabet such as $p, q, r, s, \cdots$ to denote propositions. The truth value of a proposition is true, denoted by $T$ or $1$, if it is a true proposition and false, denoted by $F$ or $0$, if it is a false proposition. The following are examples of propositions.

**EXAMPLE 1.1.1**  Each of these declarative sentences is a proposition:

1. $\sqrt{2}$ is irrational.
2. Beijing is the capital of China.
3. $1+3=5$.
4. $9+10 \leqslant 12$.
5. We'll live on the Moon by the end of this century.

Propositions 1 and 2 are true, and therefore their truth values are 1, whereas 3 and 4 are false, so their truth values are 0. The fifth sentence is also a proposition, although at present we do not know its truth value. But by the end of this century, we'll know its truth value. Some sentences that are not propositions are given in the following example.

**EXAMPLE 1.1.2**  Consider the following sentences:

1. No smoking.
2. Is there life on the Moon?
3. $x=2$.
4. $x>y$.

Sentences 1 and 2 are not propositions because they are not declarative sentences. Sentences 3 and 4 are not propositions because they are neither true nor false, since the variables in these sentences have not been assigned values.

The propositions in EXAMPLE 1.1.1 are called **primitive** propositions, for there is no way to break them down into simpler propositions. **Compound proposition**, on the other hand, is formed by applying logical operators to one or more primitive propositions. The logical operators, also called **connectives**, are denoted by the following symbols and words:

$\neg$   not, negation.
$\wedge$   and, conjunction.
$\vee$   or, disjunction.
$\rightarrow$   conditional, implication.
$\leftrightarrow$   if and only if, biconditional.

**DEFINITION 1.1.1** Let $p$ be a proposition. The negation of $p$, denoted by $\neg p$, is a new proposition. If $p$ is true, then $\neg p$ is false. If $p$ is false, then $\neg p$ is true. The proposition $\neg p$ is read "not $p$".

**EXAMPLE 1.1.3** Let $p$ denote the proposition "Three Gorges Project is located in Yichang". Then $\neg p$ is "Three Gorges Project is not located in Yichang."

The negation of a proposition can be regarded as the result of the operation of Negation Operator on a proposition. The other connectives introduced above can also be used to construct compound propositions from two or more existing propositions.

**DEFINITION 1.1.2** Let $p$ and $q$ be propositions. The compound proposition "$p$ and $q$", denoted by $p \wedge q$, is the proposition that is true when both $p$ and $q$ are true and is false otherwise. The proposition $p \wedge q$ is called the conjunction of $p$ and $q$.

**EXAMPLE 1.1.4** Translate these sentences into logical expressions.
1. Jane likes singing and dancing.
2. Ruth and Richard are classmates.
3. The sun shone on the sea and the waves danced and sparkled.
4. Jack is an intelligent child, but not diligent in his work.

*Solution*:
1. Let $p$ and $q$ represent the primitive propositions "Jane likes singing" and "Jane likes dancing", respectively. The compound proposition "Jane likes singing and dancing" can be expressed as $p \wedge q$.
2. It is a primitive proposition. We can use a letter $p$ to represent it.
3. Let $p, q$ and $r$ denote "The sun shone on the sea", "The waves danced" and "The waves sparkled", respectively. The compound proposition can be represented as $p \wedge q \wedge r$.
4. Let $p$ and $q$ denote the propositions "Jack is an intelligent child" and "Jack is diligent in his work", respectively. Then this compound proposition can be represented as $p \wedge \neg q$.

**DEFINITION 1.1.3** Let $p$ and $q$ be propositions, the expression $p \vee q$ denotes the disjunction of $p$ and $q$, and is read "$p$ or $q$". The proposition $p \vee q$ is false when $p$ and

$q$ are both false, and true otherwise.

We use the connective "or" in disjunction in the **inclusive** sense, that is, a disjunction is true when at least one of the two propositions is true.

**EXAMPLE 1.1.5** Let $p$ denote the proposition "Advanced Mathematics is a required course for college freshmen" and $q$ denote "Margaret Mitchell wrote 'Gone with the wind'". Then $p \vee q$ is "Advanced Mathematics is a required course for college freshmen, or Margaret Mitchell wrote 'Gone with the wind'."

Sometimes, we use *or* in an exclusive sense. The **exclusive or** is denoted by $\oplus$. The proposition $p \oplus q$ is true if one or the other but not both of the propositions $p$ and $q$ is true. One way to express $p \oplus q$ for $p$ and $q$ in the above example is "Advanced Mathematics is a required course for college freshmen, or Margaret Mitchell wrote 'Gone with the wind', but not both."

**DEFINITION 1.1.4** Let $p$ and $q$ be propositions. The implication $p \rightarrow q$ is the proposition which is false when $p$ is true and $q$ is false, and true otherwise. $p$ is called the antecedent, premise, or hypothesis, and $q$ is called the consequence or conclusion of $p \rightarrow q$.

An implication is sometimes called a **conditional statement**. Some common ways to read the expression $p \rightarrow q$ are "if $p$ then $q$", "$p$ implies $q$", "$p$ is sufficient for $q$", "since $p$, therefore $q$", "$p$ only if $q$", "$q$ whenever $p$", "$q$ is necessary for $p$", "$q$ follows from $p$", "$q$ if $p$", "$\neg p$ unless $q$".

**EXAMPLE 1.1.6** Denote the following propositions in symbolic forms, and find their truth values.

    1. If $2+3=5$, then snow is white.
    2. If $2+3 \neq 5$, then snow is white.
    3. If $2+3 \neq 5$, then snow is not white.
    4. If $2+3=5$, then snow is not white.
    5. 4 is a divisor of $a$ only if 2 is a divisor of $a$, where $a$ is a given positive integer.
    6. 4 is not a divisor of $a$ unless 2 is a divisor of $a$, where $a$ is a given positive integer.

*Solution*: Let $p$ and $q$ denote "$2+3=5$" and "snow is white", respectively. Then both $p$ and $q$ are true. The propositions 1, 2, 3 and 4 can be expressed as $p \rightarrow q, \neg p \rightarrow q, \neg p \rightarrow \neg q$ and $p \rightarrow \neg q$, respectively, and they have truth values 1, 1, 1, 0, respectively.

Let $r$ be "4 is a divisor of $a$", $s$ be "2 is a divisor of $a$". The proposition 5 and proposition 6 are in the form of $r \rightarrow s$. If $a$ is really divisible by 4, then $a$ is also divisible by 2. Therefore, $r \rightarrow s$ is true. If $a$ is not divisible by 4, then $r \rightarrow s$ is still true no matter whether $a$ is divisible by 2, since $r$ is false.

From EXAMPLE 1.1.6, we see that, the way we have defined implications is more general than the meaning attached to implications in the English language. For instance, proposition 5 is an implication used in normal language, since there is a relationship between the hypothesis and the conclusion. However, in proposition 1, there is no relationship between the hypothesis and conclusion. In mathematical reasoning we consider implications more general than in English.

**DEFINITION 1.1.5**  The biconditional of two propositions $p$ and $q$ is denoted by $p \leftrightarrow q$, which is read "$p$ if and only if $q$" or "$p$ is necessary and sufficient for $q$". The proposition $p \leftrightarrow q$ is true when $p$ and $q$ have the same truth values, and is false otherwise.

Sometimes "$p$ if and only if $q$" is abbreviated as "$p$ iff $q$".

**EXAMPLE 1.1.7**  Represent the following sentences in logical expressions.

1. 2 is a prime if and only if $\sqrt{5}$ is a rational number.
2. If two lines A and B are parallel, then their corresponding angles are equal, and vice versa.

*Solution*:

1. Let $p$ denote "2 is a prime" and $q$ denote "$\sqrt{5}$ is a rational number". The proposition 1 can be expressed as $p \leftrightarrow q$, its truth value is 0.
2. Let $r$ and $s$ denote "A and B are parallel" and "the corresponding angles of A and B are equal", respectively. Then the proposition 2 can be represented as $r \leftrightarrow s$, which has truth value 1.

Truth tables are used to display the relationships between the truth values of propositions, where we write "0" for false and "1" for true. As an example, Table 1.1.1 is the truth table for the propositions obtained from $p$ and $q$ by applying the six logical operators defined above.

**Table 1.1.1  The truth table for the Negation of Proposition, the Conjunction, Disjunction, Exclusive Or, Implication and Biconditional of two Propositions**

| $p$ | $q$ | $\neg p$ | $p \wedge q$ | $p \vee q$ | $p \oplus q$ | $p \rightarrow q$ | $p \leftrightarrow q$ |
|---|---|---|---|---|---|---|---|
| 0 | 0 | 1 | 0 | 0 | 0 | 1 | 1 |
| 0 | 1 | 1 | 0 | 1 | 1 | 1 | 0 |
| 1 | 0 | 0 | 0 | 1 | 1 | 0 | 0 |
| 1 | 1 | 0 | 1 | 1 | 0 | 1 | 1 |

The four possible truth assignments for $p$ and $q$ can be listed by any order. But the particular order presented above will be proved useful.

We can construct compound propositions using the logical operators defined so far. Generally parentheses are used to specify the order in which the logical operators in a compound proposition are to be applied. However, to reduce the number of parentheses, we define the hierarchy of evaluation for the connectives of the propositional calculus as: $\neg, \wedge, \vee, \rightarrow$ and $\leftrightarrow$.

**EXAMPLE 1.1.8**  Let $p, q$ and $r$ denote the propositions "Beijing is the capital of China", "Three Gorges Project is located in Yichang" and "$1+3 \neq 4$", respectively. Find the truth values for the following compound propositions:

1. $(\neg p \vee q) \wedge (\neg p \vee \neg q) \leftrightarrow r$
2. $(p \wedge r) \rightarrow (q \rightarrow \neg r)$
3. $(p \vee r) \wedge (q \vee \neg r) \rightarrow (\neg p \vee \neg q)$

*Solution*: $p, q$ and $r$ have truth values 1, 1 and 0, respectively. Consequently, proposition 1, 2 and 3 have truth values 1, 1 and 0, respectively.

## WORDS AND EXPRESSIONS

propositional logic　　　　　命题逻辑
proposition　　　　　　　　命题
truth value　　　　　　　　真值
primitive proposition　　　　原子命题

| | |
|---|---|
| compound proposition | 复合命题 |
| negation | 否定 |
| conjunction | 合取 |
| disjunction | 析取 |
| implication | 蕴涵式 |
| biconditional | 双条件式 |
| truth table | 真值表 |

# EXERCISES 1.1

1. Which of the following sentences are propositions? What are the truth values of those that are propositions?
    a. Every map in the world can be colored using four colors.
    b. Discrete Mathematics is a required course for Computer Science Major.
    c. Is the computer available?
    d. $x+3$ is a positive integer.
    e. $4+x=5$.
    f. 2 and 3 are even.

2. Let $p,q$ be primitive propositions for which the implication $p \rightarrow q$ is false. Determine the truth values of
    $$\neg p \wedge q \quad q \rightarrow p \quad \neg q \rightarrow \neg p \quad p \vee \neg q \quad q \leftrightarrow \neg p$$

3. Let $p,q,r$ denote the following statements:
    $p$: It is sunny.
    $q$: I'll go climbing.
    $r$: I'm free.
    Convert the following statements into symbolic forms.
    a. I am not free.
    b. If it is sunny, I'll go climbing.
    c. I'm free, but it is not sunny.
    d. I'll go climbing only if I am free and it is sunny.
    e. If I'm free, I will go climbing unless it is not sunny.
    f. Whenever it is sunny, I'll go climbing.
    g. Sunny day and free time are sufficient for going climbing.

4. Write the following statements in symbolic forms.
   a. It is cold and it is windy.
   b. If berries are ripe along the trail, hiking is safe if and only if grizzly bears have not been seen in the area.
   c. It is necessary to wash the boss's car to get promoted.
   d. Winds from the South imply a spring thaw.
   e. If you watch television, your mind will decay, and vice versa.
   f. Low humidity and sunshine are sufficient for me to play tennis this afternoon.
   g. It is snowing but, we will go out for a work.

5. Determine the truth value of each of the following compound propositions.
   a. If $1+1=2$, then $2+3=5$.
   b. If $1+1=3$, then $2+3=4$.
   c. If $1+1=3$, then $2+3=5$.
   d. If people can fly, then $1+2=4$.
   e. $1+1=2$ if and only if $2+3=4$.
   f. $1>2$ if and only if $3>2$.

6. There are some related implications that can be formed from $p \to q$. For example,
   The proposition $q \to p$ is called the converse of $p \to q$.
   The proposition $\neg p \to \neg q$ is called the inverse of $p \to q$.
   The proposition $\neg q \to \neg p$ is called the contra-positive of $p \to q$.
   What are the contra-positive, converse, and inverse of the implication:
   "You can ask for help whenever you need it"?

In computer programming the If-Then and If-Then-Else decision structures arise in languages such as BASIC and C++. The hypothesis $p$ is often a relational expression (such as $x>5$). This expression is a logical proposition that has truth value 0 or 1, depending on the value of the variables contained in the expression (such as $x$) at that point in the program. The conclusion $q$ may be an "executive statement" directing the program to another line or causing some results to be printed. (So $q$ is not one of the logical statements.) When dealing with "if $p$ then $q$", in this text, the computer executes $q$ only on the condition that $p$ is true. For $p$ being false, the computer goes to the next instruction in the program sequence. For the decision structure "if $p$ then $q$ else $r$", $q$ is executed when $p$ is true and $r$ is executed when $p$ is false.

7. What are the values of $m, n$ after each of these statements is encountered in a given C++ program, if $m = 3, n = 8$ before the first statement is reached? [Here the values of $m, n$ following the execution of the statement in part (a) become the values of $m, n$ for the statement in part (b), and so on, through the statement in part (g). The div operation in C++ returns the integer part of a quotient. For example 6 div 2 = 3, 5 div 2 = 2.]
   a. If $n - m == 5$, then $n = n - 2$;
   b. If $((2 * m == n)$ and $(n$ div $4 == 1))$, then $n = 4 * m - 3$;
   c. If $[(n < 8)$ or $(m$ div $2 == 2)]$, then $n = 2 * m$ else $m = 2 * n$;
   d. If $[(m < 20)$ or $(n$ div $6 == 1)]$, then $m = m - n - 5$;
   e. If $[(n == 2 * m)$ or $(n$ div $2 == 5)]$, then $m = m + 2$;
   f. If $[(n$ div $3 == 3)$ and $(m$ div $3 < > 1)]$, then $m = n$;
   g. If $m * n < > 35$ then $n = 3 * m + 7$.

Fuzzy logic is used in artificial intelligence. In fuzzy logic, a proposition has a truth value that is a number between 0 and 1, inclusive. A proposition with a truth value of 0 is false and one with a truth value of 1 is true. Truth values that are between 0 and 1 indicate varying degrees of truth. For instance, the truth value 0.99 can be assigned to the statement "Niomi is happy", since Niomi is very happy, and the truth value 0.3 can be assigned to the statement "Jack is happy", since Jack is happy slightly less than half at the time.

8. The truth value of the negation of a proposition in fuzzy logic is 1 minus the truth value of the proposition. What are the truth values of the statements "Niomi is not happy" and "Jack is not happy"?

9. The truth value of the conjunction of two propositions in fuzzy logic is the minimum of the truth values of the two propositions. What are the truth values of the statements "Niomi and Jack are happy" and "Neither Niomi nor Jack is happy"?

10. The truth value of the disjunction of two propositions in fuzzy logic is the maximum of the truth values of the two propositions. What are the truth values of the statements "Niomi is happy, or Jack is happy" and "Niomi is not happy, or Jack is not happy"?

## 1.2 Propositional WFF and Assignment

In Section 1.1, we have defined primitive proposition and compound proposition. Primitive proposition does not contain any connectives, and compound proposition contains at least one connective. For example, if $p$ and $q$ are propositions, then $\neg p$, $\neg p \vee q$, and $(p \leftrightarrow q) \vee \neg q \rightarrow p$ are compound propositions. But if $p$ and $q$ are **propositional variables**, it means that they have not been assigned any specified propositions. The expressions above then are called well-formed formula or **wff** for short, which can be pronounced as "woof". A wff is a proposition only if the propositional variables contained in the wff are assigned some specified propositions.

We define the wff by the following inductive definition for the set of propositional wffs.

### DEFINITION 1.2.1
1. A propositional variable is a wff.
2. If $A$ is a wff, then $\neg A$ is a wff.
3. If $A, B$ are wffs, then $A \wedge B, A \vee B, A \rightarrow B$ and $A \leftrightarrow B$ are wffs.

For example, $(p \rightarrow q) \wedge (q \leftrightarrow r)$, $(p \wedge q) \vee \neg r$, and $p \wedge (\neg q \vee r)$ are wffs, but $pq \rightarrow r$ and $(p \rightarrow (q \rightarrow q)$ are not.

We often use capital letters to refer to arbitrary propositional wffs. For example, if we say, $A$ is a wff, we mean that $A$ represents some arbitrary wff. We also use capital letters to denote specific propositional wffs. For example, if we want to discuss the wff $p \wedge q \wedge \neg r$ several times, we may let $W = p \wedge q \wedge \neg r$. Then we can refer to $W$ instead of always writing down the symbols $p \wedge q \wedge \neg r$.

Since there may exist some propositional variables in a wff, we usually don't know its truth value. If all propositional variables in a wff are assigned some specified propositions, the wff becomes a proposition. For example, $(p \vee q) \rightarrow r$ is a wff. Let $p$ be "2 is a prime", $q$ "3 is an even" and $r$ "5 is a rational number". Then $p$ and $r$ are true, but $q$ is false. This wff can be translated as "if 2 is a prime, or 3 is an even, then 5 is a rational number", which is true. In fact, the truth value of a wff depends on the truth values of the propositional variables contained in the wff.

According to DEFINITION 1.2.1, we can translate many sentences in natural languages (such as English) into expressions involving propositional variables and logical connectives. English (or any other human language) is often ambiguous, but the ambiguity can be removed by translating sentences into logical expressions. Moreover, once we have translated sentences from English into logical expressions, we then can manipulate them, and use rules of reference to reason them.

**EXAMPLE 1.2.1** Translate the following English sentence into a logical expression:
"You can enroll computer science only if you have taken the courses of Discrete Mathematics and Advanced Mathematics."
*Solution*: There are many ways to translate this sentence into a logical expression. Although it is possible to represent the sentence by a single propositional variable, such as $p$, this would not be useful to analyze its meaning or to do reasoning with it. Instead, we use a propositional variable to represent each sentence part and determine the appropriate logical connectives between the variables. Let $q, r$ and $s$ represent "You can enroll computer science", "You have taken the course of Discrete Mathematics", and "You have taken the course of Advanced Mathematics", respectively. Noticing that "only if" is one way to express an implication, this sentence can be represented by $q \rightarrow (r \wedge s)$.

**EXAMPLE 1.2.2** Translate the following English sentence into a logical expression:
"You cannot go to Internet cafe if you are not adult unless you are accompanied by your parents."
*Solution*: Again, we will use a propositional variable to represent each of the sentence part and to decide on the appropriate logical connectives between the variables. Let $q, r$ and $s$ represent "You can go to the Internet cafe", "You are an adult", and "You are accompanied by your parents", respectively. Then the sentence can be translated as $(\neg r \wedge \neg s) \rightarrow \neg q$, or $\neg(\neg r \rightarrow \neg q) \rightarrow s$.

Now we will discuss the truth value of a wff under all possible assignments of truth values to all propositional variables contained in it. Any wff has a unique truth table. For example, suppose we want to find the truth table for the wff $\neg p \rightarrow (q \vee r)$. Then we can construct the truth table in the following way: Begin by writing down all possible truth values for the three variables $p, q$ and $r$. This gives us a table with eight lines. Next, compute a column of values for $\neg p$. Then compute a column of values for $q \vee r$. Finally, use these two columns to compute the column of values for $\neg p \rightarrow (q \vee r)$. The final result is Table 1.2.1, from which we conclude that the wff is false when $p, q, r$ are false, and the wff is true otherwise.

**Table 1.2.1** The truth table for $\neg p \rightarrow (q \vee r)$

| $p$ | $q$ | $r$ | $\neg p$ | $q \vee r$ | $\neg p \rightarrow (q \vee r)$ |
|---|---|---|---|---|---|
| 0 | 0 | 0 | 1 | 0 | 0 |
| 0 | 0 | 1 | 1 | 1 | 1 |
| 0 | 1 | 0 | 1 | 1 | 1 |
| 0 | 1 | 1 | 1 | 1 | 1 |
| 1 | 0 | 0 | 0 | 0 | 1 |
| 1 | 0 | 1 | 0 | 1 | 1 |
| 1 | 1 | 0 | 0 | 1 | 1 |
| 1 | 1 | 1 | 0 | 1 | 1 |

**EXAMPLE 1.2.3** Find the truth table for the following wffs.

1. $q \wedge (\neg r \rightarrow p)$
2. $(p \wedge \neg q) \leftrightarrow (q \wedge \neg p)$
3. $p \rightarrow (p \vee q)$
4. $p \wedge (\neg p \wedge q)$

*Solution*: We construct the truth tables as following.

**Table 1.2.2** The truth table for $q \wedge (\neg r \rightarrow p)$

| $p$ | $q$ | $r$ | $\neg r$ | $\neg r \rightarrow p$ | $q \wedge (\neg r \rightarrow p)$ |
|---|---|---|---|---|---|
| 0 | 0 | 0 | 1 | 0 | 0 |
| 0 | 0 | 1 | 0 | 1 | 0 |
| 0 | 1 | 0 | 1 | 0 | 0 |
| 0 | 1 | 1 | 0 | 1 | 1 |
| 1 | 0 | 0 | 1 | 1 | 0 |
| 1 | 0 | 1 | 0 | 1 | 0 |
| 1 | 1 | 0 | 1 | 1 | 1 |
| 1 | 1 | 1 | 0 | 1 | 1 |

**Table 1.2.3** The truth table for $(p \wedge \neg q) \leftrightarrow (q \wedge \neg p)$

| $p$ | $q$ | $\neg q$ | $\neg p$ | $p \wedge \neg q$ | $q \wedge \neg p$ | $(p \wedge \neg q) \leftrightarrow (q \wedge \neg p)$ |
|---|---|---|---|---|---|---|
| 0 | 0 | 1 | 1 | 0 | 0 | 1 |
| 0 | 1 | 0 | 1 | 0 | 1 | 0 |
| 1 | 0 | 1 | 0 | 1 | 0 | 0 |
| 1 | 1 | 0 | 0 | 0 | 0 | 1 |

**Table 1.2.4** The truth table for $p \rightarrow (p \vee q)$

| $p$ | $q$ | $p \vee q$ | $p \rightarrow (p \vee q)$ |
|---|---|---|---|
| 0 | 0 | 0 | 1 |
| 0 | 1 | 1 | 1 |
| 1 | 0 | 1 | 1 |
| 1 | 1 | 1 | 1 |

**Table 1.2.5** The truth table for $p \wedge (\neg p \wedge q)$

| $p$ | $q$ | $\neg p$ | $\neg p \wedge q$ | $p \wedge (\neg p \wedge q)$ |
|---|---|---|---|---|
| 0 | 0 | 1 | 0 | 0 |
| 0 | 1 | 1 | 1 | 0 |
| 1 | 0 | 0 | 0 | 0 |
| 1 | 1 | 0 | 0 | 0 |

The third and fourth wffs in the example are two special types of compound propositions. The results in column 4 of Table 1.2.4 and column 5 of Table 1.2.5 reveal that the proposition $p \rightarrow (p \vee q)$ is true and the proposition $p \wedge (\neg p \wedge q)$ is false for all truth value assignments for the propositional variables $p$ and $q$.

**DEFINITION 1.2.2** A compound proposition is called a **tautology** if it is true for all truth value assignments for propositional variables that occur in it. A compound proposition that is always false is called a **contradiction**. A proposition that is neither a tautology nor a contradiction is called a **contingency**. A compound proposition is called **satisfiable** if there is an assignment of truth values to the variables in the proposition that makes the compound proposition true.

In EXAMPLE 1.2.3, wff 1 and wff 2 are contingencies, wff 3 is a tautology and wff 4 is a contradiction. Certainly, wff 1, wff 2, and wff 3 are satisfiable.

## WORDS AND EXPRESSIONS

wff ( well-formed formula )　　合式公式,命题公式
tautology　　重言式,永真式
contradiction　　永假式,矛盾式
contingency　　相依式,列联式
satisfiable　　可满足的

## EXERCISES 1.2

1. Write down the parenthesized version of each of the following expressions.

   $\neg p \wedge q \to p \vee \neg r$

   $p \vee q \wedge \neg r \to \neg p \vee r \to q$

   $p \to q \vee \neg r \wedge s \wedge t \to \neg q$

2. Construct a truth table for each of the following compound propositions.

   $\neg(p \vee \neg q) \to \neg p$　　　$p \to (q \to r)$　　　$(p \to q) \to r$

   $(p \to q) \to (q \to p)$　　$(p \wedge (p \to q)) \to q$　　$(p \wedge q) \to p$

   $q \leftrightarrow (\neg p \vee \neg r)$　　$((p \to q) \wedge (q \to r)) \to (p \to r)$

3. Determine all the truth value assignments, if any, for the propositional variables $p$, $q$, $r$, $s$ and $t$ that make each of the following compound propositions false.

   $(p \wedge q) \wedge r \to s \vee t$　　　$p \wedge (\neg q \vee r) \to \neg s \wedge \neg t$

4. Answer the following questions:

   a. How many rows are needed for the truth table of the compound proposition $(\neg p \wedge q) \leftrightarrow ((r \vee s) \to t)$?

   b. Suppose $A$ is a wff containing $n$ primitive propositions. How many rows are needed to construct the truth table for $A$?

5. Which of the compound propositions in Question 2 are satisfiable?

6. Verify that $(p \to (q \to r)) \to ((p \to q) \to (p \to r))$ is a tautology.

7. Determine whether $(\neg p \wedge (p \to q)) \to \neg q$ is a tautology.

8. Determine whether $(\neg p \wedge (p \to q)) \to \neg p$ is a tautology.

9. Show that the following wffs are tautologies.

a. $p \wedge (p \rightarrow q) \rightarrow q$
b. $(p \rightarrow q) \wedge (q \rightarrow r) \rightarrow (p \rightarrow r)$
c. $(p \rightarrow q) \wedge \neg q \rightarrow \neg p$
d. $(p \vee q) \wedge \neg p \rightarrow q$
e. $p \wedge q \rightarrow p$
f. $p \rightarrow (p \vee q)$
g. $(p \rightarrow r) \wedge (q \rightarrow r) \rightarrow ((p \vee q) \rightarrow r)$
h. $(p \rightarrow q) \wedge (r \rightarrow s) \wedge (p \vee r) \rightarrow (q \vee s)$
i. $(p \rightarrow q) \wedge (r \rightarrow s) \wedge (\neg q \vee \neg s) \rightarrow (\neg p \vee \neg r)$

## 1.3 Propositional Equivalences

An important method used extensively in the construction of mathematical arguments is to replace a statement by another statement with the same truth value. This is actually a method that produces propositions with the same truth value as a given proposition. The methodology is based on the fact that some wffs have the same truth values in all possible cases even though their expressions are different. For example, the wffs $p \rightarrow q$ and $\neg p \vee q$ have the same truth values in all possible cases because they have the same truth table. Two wffs are said to be **logically equivalent** if they have the same truth table. We formally define this notion below.

**DEFINITION 1.3.1** Two wffs $A$ and $B$ are called logically equivalent, denoted by $A \equiv B$, if the wff $A \leftrightarrow B$ is a tautology.

**EXAMPLE 1.3.1** (Distributive law of disjunction over conjunction) The wffs $p \vee (q \wedge r)$ and $(p \vee q) \wedge (p \vee r)$ are logically equivalent.

*Proof*: We construct the truth table for these wffs in Table 1.3.1. Since the truth values of $p \vee (q \wedge r)$ and $(p \vee q) \wedge (p \vee r)$ agree, these wffs are logically equivalent.

**Table 1.3.1** Truth values of $p \vee (q \wedge r)$ and $(p \vee q) \wedge (p \vee r)$

| $p$ | $q$ | $r$ | $p \vee (q \wedge r)$ | $p \vee q$ | $p \vee r$ | $(p \vee q) \wedge (p \vee r)$ |
|---|---|---|---|---|---|---|
| 0 | 0 | 0 | 0 | 0 | 0 | 0 |
| 0 | 0 | 1 | 0 | 0 | 1 | 0 |
| 0 | 1 | 0 | 0 | 1 | 0 | 0 |
| 0 | 1 | 1 | 1 | 1 | 1 | 1 |

**Continue**

| p | q | r | $p \lor (q \land r)$ | $p \lor q$ | $p \lor r$ | $(p \lor q) \land (p \lor r)$ |
|---|---|---|---|---|---|---|
| 1 | 0 | 0 | 1 | 1 | 1 | 1 |
| 1 | 0 | 1 | 1 | 1 | 1 | 1 |
| 1 | 1 | 0 | 1 | 1 | 1 | 1 |
| 1 | 1 | 1 | 1 | 1 | 1 | 1 |

**EXAMPLE 1.3.2** Show that $\neg(p \lor q)$ and $\neg p \land \neg q$ are logically equivalent. (This equivalence is one of **De Morgan's laws** for propositions.)

*Proof*: We construct Table 1.3.2 for $\neg(p \lor q) \leftrightarrow \neg p \land \neg q$. From the table, this compound proposition is a tautology. Hence $\neg(p \lor q) \equiv \neg p \land \neg q$.

**Table 1.3.2** Truth values of $\neg(p \lor q) \leftrightarrow \neg p \land \neg q$

| p | q | $\neg(p \lor q)$ | $\neg p \land \neg q$ | $\neg(p \lor q) \leftrightarrow \neg p \land \neg q$ |
|---|---|---|---|---|
| 0 | 0 | 1 | 1 | 1 |
| 0 | 1 | 0 | 0 | 1 |
| 1 | 0 | 0 | 0 | 1 |
| 1 | 1 | 0 | 0 | 1 |

The associative law for disjunction shows that the expression $p \lor q \lor r$ is well defined, in the sense that it does not matter whether we first take the disjunction of $p$ and $q$ and then the disjunction of $p \lor q$ and $r$, or we first take the disjunction of $q$ and $r$ and then take the disjunction of $p$ and $q \lor r$. Similarly the expression $p \land q \land r$ is also well defined. By extending this reasoning, it is easy to see that $p_1 \lor p_2 \lor \cdots \lor p_n$ and $p_1 \land p_2 \land \cdots \land p_n$ are well defined whenever $p_1, p_2, \cdots, p_n$ are propositions. Furthermore, De Morgan's laws can be extend to

$$\neg(p_1 \lor p_2 \lor \cdots \lor p_n) \equiv \neg p_1 \land \neg p_2 \land \cdots \land \neg p_n \text{ and}$$
$$\neg(p_1 \land p_2 \land \cdots \land p_n) \equiv \neg p_1 \lor \neg p_2 \lor \cdots \lor \neg p_n.$$

A truth table can be used to determine whether a wff is a tautology. This can be done by hand for a wff with a small number of variables. But when the number of variables grows, it becomes impractical. For instance there are $2^{20} = 1,048,576$ rows in the truth table for a proposition with 20 variables. In this occasion a computer is needed to help to determine whether a wff in 20 variables is a tautology. But when there are 1000 variables, can even a computer determine in a reasonable amount of time whether a wff is a tautology? Checking every one of the $2^{1000}$ (a number with more than 300 decimal digits) possible combinations of truth values simply cannot be done by a computer in many years. At present there are no other known procedures that a computer can follow to determine in a reasonable amount of time whether a wff in such a large number of variables is a tautology. This is the place where logical equivalences are needed.

Using the concepts of logical equivalence, tautology, and contradiction, we list the following important equivalences. In these equivalences, $T$ denotes a proposition that is always true and $F$ denotes a proposition that is always false. It is easy to verify all these equivalences by truth tables, so they are left as exercises. In the following, $A$, $B$ and $C$ denote arbitrary wffs.

## Logical Basic Equivalences

1. Double negation law
   $\neg(\neg A) \equiv A$
2. Idempotent laws
   $A \vee A \equiv A$
   $A \wedge A \equiv A$
3. Associative laws
   $(A \vee B) \vee C \equiv A \vee (B \vee C)$
   $(A \wedge B) \wedge C \equiv A \wedge (B \wedge C)$
4. Distributive laws
   $A \vee (B \wedge C) \equiv (A \vee B) \wedge (A \vee C)$
   $A \wedge (B \vee C) \equiv (A \wedge B) \vee (A \wedge C)$
5. De Morgan's laws
   $\neg(A \wedge B) \equiv \neg A \vee \neg B$
   $\neg(A \vee B) \equiv \neg A \wedge \neg B$
6. Identity laws
   $A \wedge T \equiv A$
   $A \vee F \equiv A$
7. Domination laws
   $A \vee T \equiv T$
   $A \wedge F \equiv F$
8. Negation laws
   $A \vee \neg A \equiv T$

$A \wedge \neg A \equiv F$

9. Absorption laws

$A \wedge (A \vee B) \equiv A$

$A \vee (A \wedge B) \equiv A$

10. Logical Equivalences involving implication

$A \rightarrow B \equiv \neg A \vee B$

$A \rightarrow B \equiv \neg B \rightarrow \neg A$

$(A \rightarrow B) \wedge (A \rightarrow C) \equiv A \rightarrow (B \wedge C)$

$(A \rightarrow C) \wedge (B \rightarrow C) \equiv (A \vee B) \rightarrow C$

$(A \rightarrow B) \vee (A \rightarrow C) \equiv A \rightarrow (B \vee C)$

$(A \rightarrow C) \vee (B \rightarrow C) \equiv (A \wedge B) \rightarrow C$

$(A \rightarrow B) \wedge (A \rightarrow \neg B) \equiv \neg A$

11. Logical Equivalences involving biconditional

$A \leftrightarrow B \equiv (A \rightarrow B) \wedge (B \rightarrow A)$

$A \leftrightarrow B \equiv \neg A \leftrightarrow \neg B$

$A \leftrightarrow B \equiv (A \wedge B) \vee (\neg A \wedge \neg B)$

$\neg (A \leftrightarrow B) \equiv A \leftrightarrow \neg B$

We can use these equivalences to show the equivalences of two wffs without checking truth tables. But first we need to observe two general properties of equivalences.

The first thing to observe is that equivalences have transitive property which can be stated as the following:

**PROPERTY 1.3.1** Let $A$, $B$ and $C$ be any wffs. If $A \equiv B$ and $B \equiv C$, then $A \equiv C$.

PROPERTY 1.3.1 allows us to write a sequence of equivalences and then conclude that the first wff is equivalent to the last one.

The next thing to observe is the **replacement rule** of equivalences, which can be considered after we have defined the concept of subwff.

**DEFINITION 1.3.2** A wff $X$ is called a **subwff** of $A$ if $X$ is part of wff $A$.

For example, let $A = (p \rightarrow q) \wedge r$, then $X = p \rightarrow q$ is a subwff of $A$.

**PROPERTY 1.3.2** (Replacement rule) If $\Phi(A)$ is a wff with $A$ as its subwff, and $A \equiv B$. Then $\Phi(A) \equiv \Phi(B)$.

For example, suppose we want to simplify the wff $B \to (A \lor (A \land B))$. We notice that the absorption law gives $A \lor (A \land B) \equiv A$. Therefore, we can apply the replacement law and write the equivalence: $B \to (A \lor (A \land B)) \equiv B \to A$.

Let's use some examples to illustrate the process of showing that two wffs are equivalent without checking their truth tables.

**EXAMPLE 1.3.3** Show that $\neg(p \lor (\neg p \land q))$ and $\neg p \land \neg q$ are logically equivalent.

*Proof:* We will prove it using equivalences that we already know.

$\quad \neg(p \lor (\neg p \land q))$
$\equiv \neg p \land \neg(\neg p \land q)$     (De Morgan's law)
$\equiv \neg p \land (p \lor \neg q)$     (De Morgan's law)
$\equiv (\neg p \land p) \lor (\neg p \land \neg q)$     (Distributive law of $\land$ over $\lor$)
$\equiv F \lor (\neg p \land \neg q)$     (Negation law)
$\equiv \neg p \land \neg q$     (Identity law)

Consequently, $\neg(p \lor (\neg p \land q))$ and $\neg p \land \neg q$ are logically equivalent.

**EXAMPLE 1.3.4** Show that $p \to (q \to r) \equiv q \to (p \to r)$.

*Proof:* $p \to (q \to r) \equiv p \to (\neg q \lor r)$     ($\because q \to r \equiv \neg q \lor r$)
$\quad\quad\quad\quad\quad\quad \equiv \neg p \lor (\neg q \lor r)$     ($\because A \to B \equiv \neg A \lor B$)
$\quad\quad\quad\quad\quad\quad \equiv \neg q \lor (\neg p \lor r)$     (Associative law)
$\quad\quad\quad\quad\quad\quad \equiv q \to (\neg p \lor r)$     ($\neg A \lor B \equiv A \to B$)
$\quad\quad\quad\quad\quad\quad \equiv q \to (p \to r)$     ($\neg A \lor B \equiv A \to B$)

Therefore, $p \to (q \to r) \equiv q \to (p \to r)$.

**EXAMPLE 1.3.5** Which types are these wffs?

1. $(p \to q) \land p \to q$    2. $\neg(p \to (p \lor q)) \land r$    3. $p \land (((p \lor q) \land \neg p) \to q)$

*Solution:* In order to determine the types of these wffs, we need to simplify these wffs using equivalences.

$\quad$ 1. $(p \to q) \land p \to q \equiv (\neg p \lor q) \land p \to q$
$\quad\quad\quad\quad\quad\quad\quad\quad \equiv \neg((\neg p \lor q) \land p) \lor q$
$\quad\quad\quad\quad\quad\quad\quad\quad \equiv \neg(\neg p \lor q) \lor \neg p \lor q$
$\quad\quad\quad\quad\quad\quad\quad\quad \equiv (p \land \neg q) \lor \neg p \lor q$
$\quad\quad\quad\quad\quad\quad\quad\quad \equiv ((p \lor \neg p) \land (\neg q \lor \neg p)) \lor q$
$\quad\quad\quad\quad\quad\quad\quad\quad \equiv (\neg p \lor \neg q) \lor q$
$\quad\quad\quad\quad\quad\quad\quad\quad \equiv \neg p \lor (q \lor \neg q)$

$$\equiv \neg p \vee T$$
$$\equiv T$$

The wff $(p \rightarrow q) \wedge p \rightarrow q$ is a tautology.

2.  $\neg(p \rightarrow (p \vee q)) \wedge r \equiv \neg(\neg p \vee (p \vee q)) \wedge r$
$$\equiv \neg((\neg p \vee p) \vee q) \wedge r$$
$$\equiv \neg T \wedge r$$
$$\equiv F \wedge r$$
$$\equiv F$$

The wff $\neg(p \rightarrow (p \vee q)) \wedge r$ is a contradiction.

3.  $p \wedge (((p \vee q) \wedge \neg p) \rightarrow q) \equiv p \wedge (\neg((p \vee q) \wedge \neg p) \vee q)$
$$\equiv p \wedge (\neg((p \wedge \neg p) \vee (q \wedge \neg p)) \vee q)$$
$$\equiv p \wedge (\neg(F \vee (q \wedge \neg p)) \vee q)$$
$$\equiv p \wedge (\neg q \vee p \vee q)$$
$$\equiv p \wedge T$$
$$\equiv p$$

The wff $p \wedge (((p \vee q) \wedge \neg p) \rightarrow q)$ is false only if $p$ is false, otherwise, it is true.

**EXAMPLE 1.3.6** Show that $(p \wedge q) \rightarrow (p \vee q)$ is a tautology.

*Proof*: To show this wff is a tautology, we use logical equivalences to demonstrate that it is logically equivalent to $T$. (Note this could also be done using a truth table.)

$$(p \wedge q) \rightarrow (p \vee q) \equiv \neg(p \wedge q) \vee (p \vee q)$$
$$\equiv (\neg p \vee \neg q) \vee p \vee q$$
$$\equiv (\neg p \vee p) \vee (\neg q \vee q)$$
$$\equiv T$$

A switching network is made up of wires and switches connecting two terminals $T_1$ and $T_2$. In such a network, any switch is either open (0), so that no current flows through it, or closed (1), so that current does flow through it.

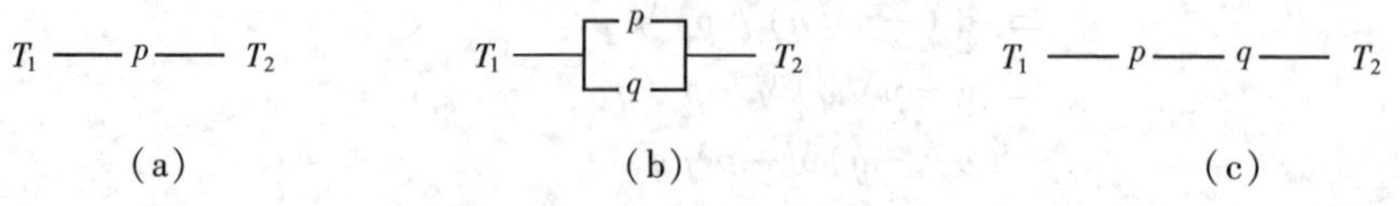

```
T₁ ——p—— T₂        T₁ ——⌈—p—⌉—— T₂        T₁ ——p——q—— T₂
                       ⌊—q—⌋

     (a)                   (b)                   (c)
```

**Figure 1.3.1  Switching Networks**

In Figure 1.3.1 we have in part (a) a network with one switch. Each of parts (b) and

(c) contains two independent switches. For the network in part (b), current flows from $T_1$ to $T_2$ if either of the switches $p$ or $q$ is closed. We call this a **parallel network** and represent it by $p \lor q$. The network in part (c) requires that both of the switches $p$ and $q$ are closed in order for current to flow from $T_1$ to $T_2$. Here the switches are in **series**, and this network is represented by $p \land q$.

The switches in a network don't need to act independently. In Figure 1.3.2(a), the switches labeled $s$ and $\neg s$ are not independent. $s$ is open (closed) if and only if $\neg s$ is simultaneously closed (open). The same is true for the switches $q$ and $\neg q$.

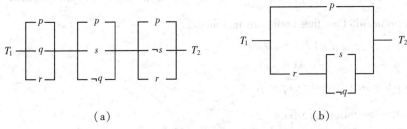

(a)  (b)

**Figure 1.3.2   Networks having dependent variables**

The network in Figure 1.3.2 (a) is represented by the expression:
$$(p \lor q \lor r) \land (p \lor s \lor \neg q) \land (p \lor \neg s \lor r).$$
We can simplify this expression as follows:
$$(p \lor q \lor r) \land (p \lor s \lor \neg q) \land (p \lor \neg s \lor r)$$
$$\equiv p \lor ((q \lor r) \land (s \lor \neg q) \land (\neg s \lor r))$$
$$\equiv p \lor ((s \lor \neg q) \land ((q \land \neg s) \lor r))$$
$$\equiv p \lor (\neg(\neg s \land q) \land ((q \land \neg s) \lor r))$$
$$\equiv p \lor ((s \lor \neg q) \land r).$$

Hence the network shown in Figure 1.3.2 (b) is equivalent to the original network in the sense that the current flows from $T_1$ to $T_2$ in network (a) exactly when it does so in network (b). But in (b) the network has only four switches, five fewer than those in the original network.

## WORDS AND EXPRESSIONS

| | |
|---|---|
| double negation | 双重否定 |
| idempotent law | 幂等律 |
| associative law | 结合律 |

distributive law     分配律
identity law         同一律
domination law       零律
negation law         否定律
absorption law       吸收律
replacement rule     替换法则

# EXERCISES 1.3

1. Use truth tables to verify the basic logical equivalences given in this section.

2. Determine whether these wffs are tautologies.
   a. $(p \vee q) \rightarrow (r \vee p) \wedge (\neg r \vee q)$
   b. $(p \rightarrow q) \wedge (q \rightarrow r) \rightarrow (r \rightarrow p)$
   c. $(p \vee q \rightarrow r) \wedge p \rightarrow (r \rightarrow q)$

3. Simplify the following wffs.
   a. $p \vee (\neg p \vee (q \wedge \neg q))$
   b. $(p \wedge q \wedge r) \vee (\neg p \wedge q \wedge r)$
   c. $\neg (p \vee (\neg p \wedge q))$

4. Verify each of the following equivalences using basic equivalences.
   a. $(p \rightarrow q) \wedge (p \vee q) \equiv q$
   b. $(p \wedge q) \rightarrow r \equiv p \rightarrow (q \rightarrow r)$
   c. $p \rightarrow (q \vee r) \equiv \neg r \rightarrow (p \rightarrow q)$
   d. $((p \wedge q \wedge r) \rightarrow s) \wedge (r \rightarrow (p \vee q \vee s)) \equiv r \wedge (p \leftrightarrow q) \rightarrow s$
   e. $((p \wedge q) \rightarrow r) \wedge (q \rightarrow (s \vee r)) \equiv q \wedge (s \rightarrow p) \rightarrow r$

Take one more look at the Logical Basic Equivalences. Those laws from the second through the 9$^{th}$ fall naturally into pairs. This pairing idea will help us discussing the following concept.

The **dual** of a compound proposition that contains only the logical connectives $\vee$, $\wedge$ and $\neg$ is the proposition obtained by replacing each $\vee$ by $\wedge$, each $\wedge$ by $\vee$, each $T$ by $F$, and each $F$ by $T$. The dual of $W$ is denoted by $W^*$.

5. Suppose $W^*$ is the dual of $W$, $p_1, p_2, \cdots, p_n$ are all propositional variables occurring in $W$. Show that the following two equivalences hold.

$$\neg W(p_1, p_2, \cdots, p_n) \equiv W^*(\neg p_1, \neg p_2, \cdots, \neg p_n)$$
$$W(\neg p_1, \neg p_2, \cdots, \neg p_n) \equiv \neg W^*(p_1, p_2, \cdots, p_n)$$

(Hint: use De Morgan's laws)

6. Suppose that wffs $A$ and $B$ have the same propositional variables, show that if $A \equiv B$ then $A^* \equiv B^*$ (Hint: let $p_1, p_2, \cdots, p_n$ be all propositional variables of $A$ and $B$, if $A \leftrightarrow B$ is a tautology, then $A(\neg p_1, \neg p_2, \cdots, \neg p_n) \leftrightarrow B(\neg p_1, \neg p_2, \cdots, \neg p_n)$ is a tautology. Exercise 5 implies the conclusion.)

7. Find the dual of each of these wffs.
   a. $\neg p \wedge q \wedge r$
   b. $(p \wedge q \wedge r) \vee s$
   c. $(p \vee F) \wedge (q \vee T)$

## 1.4  Disjunctive Normal Form

This section will discuss some useful forms for propositional wffs containing $n$ propositional variables. Let's start with some definitions.

**DEFINITION 1.4.1**  A propositional variable or its negation is called a **literal**. A **fundamental conjunction** is either a literal or a conjunction of two or more literals.

For example, $p, \neg q$ are literals and $\neg p, q, p \wedge q, p \wedge \neg q \wedge r$ are fundamental conjunctions.

**DEFINITION 1.4.2**  A **disjunctive normal form** is either a fundamental conjunction or a disjunction of two or more fundamental conjunctions.

From DEFINITION 1.4.2 it is easy to see that a wff $W$ is a disjunctive normal form if and only if $W = A_1 \vee A_2 \vee \cdots \vee A_k$ where every $A_i$ ($1 \leq i \leq k$) is a fundamental conjunction.

For example, $p \wedge q, p \vee (p \wedge q)$ and $(p \wedge \neg q) \vee q \vee (\neg q \wedge r)$ are disjunctive normal forms.

For any wff $W$ we can construct an equivalent disjunctive normal form by three steps:

**Step 1.**  Remove all occurrences (if there are any) of the connectives $\rightarrow$ and $\leftrightarrow$ by using the equivalences $A \leftrightarrow B \equiv (A \wedge B) \vee (\neg A \wedge \neg B)$ and $A \rightarrow B \equiv \neg A \vee B$ such that the wff contains only connectives of $\land, \neg$ and $\lor$.

**Step 2.** Move all negations inside to create literals by using De Morgan's equivalences
$$\neg(A \land B) \equiv \neg A \lor \neg B \text{ and } \neg(A \lor B) \equiv \neg A \land \neg B$$

**Step 3.** Apply the distributive law and associative law to obtain a disjunctive normal form.

**EXAMPLE 1.4.1** Construct a disjunctive normal form for the wff $(p \to q) \leftrightarrow r$.

**Solution:** 
$$(p \to q) \leftrightarrow r \equiv (\neg p \lor q) \leftrightarrow r$$
$$\equiv ((\neg p \lor q) \land r) \lor (\neg(\neg p \lor q) \land \neg r)$$
$$\equiv (\neg p \land r) \lor (q \land r) \lor ((p \land \neg q) \land \neg r)$$
$$\equiv (\neg p \land r) \lor (q \land r) \lor (p \land \neg q \land \neg r)$$

A wff may have several equivalent disjunctive normal forms. For example, the disjunctive normal form $p \lor (q \land r)$ can also be written as

$$p \lor (q \land r) \equiv (p \lor q) \land (p \lor r) \equiv (p \land p) \lor (p \land r) \lor (p \land q) \lor (q \land r).$$

But associated with any wff we can construct a unique standard disjunctive normal form called full disjunctive normal form.

**DEFINITION 1.4.3** A disjunctive normal form $W$ containing $n$ propositional variables is called a **full disjunctive normal** form if each fundamental conjunction of $W$ has exactly $n$ literals, one for each of the $n$ variables occurring in $W$.

For example, $(p \land q \land \neg r) \lor (\neg p \land q \land r) \lor (p \land \neg q \land r)$ is a full disjunctive normal form. $(p \land q \land \neg r) \lor (\neg p \land r)$ is a disjunctive normal form but not a full disjunctive normal form, because the variable $q$ does not occur in the second fundamental conjunction.

A truth table can be used to construct a full disjunctive normal form. If a wff $W$ contains $n$ propositional variables, then in the truth table of $W$, there are $2^n$ different assignments of truth values to these variables, which correspond to $2^n$ different fundamental conjunctions denoted by $m_i$, $0 \leq i \leq 2^n - 1$. If some variable $p$ is assigned 1, then put $p$ in the fundamental conjunction. If $p$ is assigned 0, then put $\neg p$ in the fundamental conjunction. Each integer $i$, $0 \leq i \leq 2^n - 1$ corresponds to a combination of truth values representing $m_i$. For convenience, we usually use the combination of truth values instead of the integer $i$. The full disjunctive normal form of a wff can be obtained by taking disjunctions of those $m_i$, $0 \leq i \leq 2^n - 1$, which correspond to the combinations of values for which the wff is true. Table 1.4.1 lists the fundamental conjunctions with three propositional variables.

**Table 1.4.1  The eight fundamental conjunctions with three propositional variables**

| $p$ | $q$ | $r$ | $\neg p \wedge \neg q \wedge \neg r$ | $\neg p \wedge \neg q \wedge r$ | $\neg p \wedge q \wedge \neg r$ | $\neg p \wedge q \wedge r$ |
|---|---|---|---|---|---|---|
| 0 | 0 | 0 | 1 | 0 | 0 | 0 |
| 0 | 0 | 1 | 0 | 1 | 0 | 0 |
| 0 | 1 | 0 | 0 | 0 | 1 | 0 |
| 0 | 1 | 1 | 0 | 0 | 0 | 1 |
| 1 | 0 | 0 | 0 | 0 | 0 | 0 |
| 1 | 0 | 1 | 0 | 0 | 0 | 0 |
| 1 | 1 | 0 | 0 | 0 | 0 | 0 |
| 1 | 1 | 1 | 0 | 0 | 0 | 0 |

| $p$ | $q$ | $r$ | $p \wedge \neg q \wedge \neg r$ | $p \wedge \neg q \wedge r$ | $p \wedge q \wedge \neg r$ | $p \wedge q \wedge r$ |
|---|---|---|---|---|---|---|
| 0 | 0 | 0 | 0 | 0 | 0 | 0 |
| 0 | 0 | 1 | 0 | 0 | 0 | 0 |
| 0 | 1 | 0 | 0 | 0 | 0 | 0 |
| 0 | 1 | 1 | 0 | 0 | 0 | 0 |
| 1 | 0 | 0 | 1 | 0 | 0 | 0 |
| 1 | 0 | 1 | 0 | 1 | 0 | 0 |
| 1 | 1 | 0 | 0 | 0 | 1 | 0 |
| 1 | 1 | 1 | 0 | 0 | 0 | 1 |

Using the notation of $m_i$ above, we have the following expressions from Table 1.4.1:

$m_0 \equiv \neg p \wedge \neg q \wedge \neg r$,   $m_1 \equiv \neg p \wedge \neg q \wedge r$,   $m_2 \equiv \neg p \wedge q \wedge \neg r$,
$m_3 \equiv \neg p \wedge q \wedge r$,   $m_4 \equiv p \wedge \neg q \wedge \neg r$,   $m_5 \equiv p \wedge \neg q \wedge r$,
$m_6 \equiv p \wedge q \wedge \neg r$,   $m_7 \equiv p \wedge q \wedge r$.

But for convenience, we use the following expressions instead:

$m_{000} \equiv \neg p \wedge \neg q \wedge \neg r$,   $m_{001} \equiv \neg p \wedge \neg q \wedge r$,   $m_{010} \equiv \neg p \wedge q \wedge \neg r$,
$m_{011} \equiv \neg p \wedge q \wedge r$,   $m_{100} \equiv p \wedge \neg q \wedge \neg r$,   $m_{101} \equiv p \wedge \neg q \wedge r$,
$m_{110} \equiv p \wedge q \wedge \neg r$,   $m_{111} \equiv p \wedge q \wedge r$.

**EXAMPLE 1.4.2**  Use truth tables to write the full disjunctive normal forms for the wffs $p \rightarrow q$, $p \vee q$ and $\neg (p \wedge q)$.

*Solution*: Table 1.4.2 displays the truth tables of wffs $p \rightarrow q$, $p \vee q$ and $\neg (p \wedge q)$.

**Table 1.4.2** The truth tables of wffs $p \rightarrow q$, $p \vee q$ and $\neg(p \wedge q)$

| $p$ | $q$ | $p \rightarrow q$ | $p \vee q$ | $\neg(p \wedge q)$ |
|---|---|---|---|---|
| 0 | 0 | 1 | 0 | 1 |
| 0 | 1 | 1 | 1 | 1 |
| 1 | 0 | 0 | 1 | 1 |
| 1 | 1 | 1 | 1 | 0 |

From the tables, the full disjunctive normal forms can be written as

$$p \rightarrow q \equiv (\neg p \wedge \neg q) \vee (\neg p \wedge q) \vee (p \wedge q)$$
$$p \vee q \equiv (\neg p \wedge q) \vee (p \wedge \neg q) \vee (p \wedge q)$$
$$\neg(p \wedge q) \equiv (\neg p \wedge \neg q) \vee (\neg p \wedge q) \vee (p \wedge \neg q)$$

or

$$p \rightarrow q \equiv m_0 \vee m_1 \vee m_3$$
$$p \vee q \equiv m_1 \vee m_2 \vee m_3$$
$$\neg(p \wedge q) \equiv m_0 \vee m_1 \vee m_2$$

or

$$p \rightarrow q \equiv m_{00} \vee m_{01} \vee m_{11}$$
$$p \vee q \equiv m_{01} \vee m_{10} \vee m_{11}$$
$$\neg(p \wedge q) \equiv m_{00} \vee m_{01} \vee m_{10}.$$

**EXAMPLE 1.4.3** $W$ is a wff with the truth table in Table 1.4.3, find its full disjunctive normal form.

**Table 1.4.3** The truth table for $W$

| $p$ | $q$ | $r$ | $W$ |
|---|---|---|---|
| 0 | 0 | 0 | 1 |
| 0 | 0 | 1 | 0 |
| 0 | 1 | 0 | 0 |
| 0 | 1 | 1 | 1 |
| 1 | 0 | 0 | 1 |
| 1 | 0 | 1 | 0 |
| 1 | 1 | 0 | 1 |
| 1 | 1 | 1 | 0 |

*Solution*: Using the truth table, we construct $W$ by taking disjunction of some fundamental conjunctions, which correspond to the combinations of values for which the wff $W$ is true.

$$W \equiv (\neg p \wedge \neg q \wedge \neg r) \vee (\neg p \wedge q \wedge r) \vee (p \wedge \neg q \wedge \neg r) \vee (p \wedge q \wedge \neg r)$$
$$\equiv m_0 \vee m_3 \vee m_4 \vee m_6$$

Apart from the truth table technique, we can also construct full disjunctive normal forms for wffs using equivalences. To find a full disjunctive normal form in this way, we first convert the wff to a disjunctive normal form by the usual methods: eliminate conditionals and biconditionals, move negations inside, and distribute $\wedge$ over $\vee$. Then add the extra propositional variables that do not occur in the fundamental conjunctions and use the distributive laws of $\wedge$ over $\vee$. Finally the use idempotent laws to the expression.

**EXAMPLE 1.4.4** Find the full disjunctive normal forms of wffs.
1. $(p \wedge q) \vee (\neg p \wedge r) \vee (q \wedge r)$
2. $p \to ((p \to q) \wedge \neg (\neg p \vee \neg r))$

*Solution*:

1. $(p \wedge q) \vee (\neg p \wedge r) \vee (q \wedge r)$
$\equiv ((p \wedge q) \wedge T) \vee (\neg p \wedge T \wedge r) \vee (T \wedge q \wedge r)$
$\equiv ((p \wedge q) \wedge (r \vee \neg r)) \vee (\neg p \wedge (q \vee \neg q) \wedge r) \vee ((p \vee \neg p) \wedge q \wedge r)$
$\equiv (p \wedge q \wedge r) \vee (p \wedge q \wedge \neg r) \vee (\neg p \wedge q \wedge r) \vee (\neg p \wedge \neg q \wedge r) \vee$
$(p \wedge q \wedge r) \vee (\neg p \wedge q \wedge r)$
$\equiv (p \wedge q \wedge r) \vee (p \wedge q \wedge \neg r) \vee (\neg p \wedge q \wedge r) \vee (\neg p \wedge \neg q \wedge r)$

2. $p \to ((p \to q) \wedge \neg (\neg p \vee \neg r))$
$\equiv \neg p \vee ((\neg p \vee q) \wedge (p \wedge r))$
$\equiv \neg p \vee (\neg p \wedge p \wedge r) \vee (p \wedge q \wedge r)$
$\equiv (\neg p \wedge T \wedge T) \vee F \vee (p \wedge q \wedge r)$
$\equiv (\neg p \wedge (q \vee \neg q) \wedge (r \vee \neg r)) \vee (p \wedge q \wedge r)$
$\equiv (\neg p \wedge q \wedge r) \vee (\neg p \wedge q \wedge \neg r) \vee (\neg p \wedge \neg q \wedge r) \vee (\neg p \wedge \neg q \wedge \neg r) \vee$
$(p \wedge q \wedge r)$

In a manner entirely analogous to the above discussion we can define a **fundamental disjunction** to be either a literal or the disjunction of two or more literals. A **conjunctive normal form** is either a fundamental disjunction or a conjunction of two or more fundamental disjunctions. For example, $p, \neg p \wedge (q \vee r)$ and $(p \vee q) \wedge (\neg q \vee r)$ are conjunctive normal forms.

Suppose a wff $W$ has $n$ propositional variables. A conjunctive normal form is called a

**full conjunctive normal form** if each fundamental disjunction has exactly $n$ literals, one for each of the $n$ variables that occur in $W$. For example, the following wff is a full conjunctive normal form:

$$(\neg p \vee q \vee r) \wedge (\neg p \vee \neg q \vee r).$$

There are also two ways to obtain a full conjunctive normal form for a wff $W$. One is using truth tables and the other using equivalences.

Suppose wff $W$ has $n$ propositional variables. In the truth table of $W$ there are $2^n$ different assignments of truth values corresponding to $2^n$ fundamental disjunctions, denoted by $M_i$, $0 \leqslant i \leqslant 2^n - 1$. In each $M_i$, if some variable $p$ is assigned 1, then put $\neg p$ in the disjunction. If $p$ is assigned 0, then put $p$ in the disjunction. The full conjunctive normal form of a wff can be obtained by taking conjunctions of those $M_i$, $0 \leqslant i \leqslant 2^n - 1$, which correspond to the combinations of values for which the wff is false. Table 1.4.4 lists the fundamental disjunctions with three propositional variables.

**Table 1.4.4  The list of all eight fundamental disjunctions with three propositional variables**

| $p$ | $q$ | $r$ | $p \vee q \vee r$ | $p \vee q \vee \neg r$ | $p \vee \neg q \vee r$ | $p \vee \neg q \vee \neg r$ | $\neg p \vee q \vee r$ | $\neg p \vee q \vee \neg r$ | $\neg p \vee \neg q \vee r$ | $\neg p \vee \neg q \vee \neg r$ |
|---|---|---|---|---|---|---|---|---|---|---|
| 0 | 0 | 0 | 0 | 1 | 1 | 1 | 1 | 1 | 1 | 1 |
| 0 | 0 | 1 | 1 | 0 | 1 | 1 | 1 | 1 | 1 | 1 |
| 0 | 1 | 0 | 1 | 1 | 0 | 1 | 1 | 1 | 1 | 1 |
| 0 | 1 | 1 | 1 | 1 | 1 | 0 | 1 | 1 | 1 | 1 |
| 1 | 0 | 0 | 1 | 1 | 1 | 1 | 0 | 1 | 1 | 1 |
| 1 | 0 | 1 | 1 | 1 | 1 | 1 | 1 | 0 | 1 | 1 |
| 1 | 1 | 0 | 1 | 1 | 1 | 1 | 1 | 1 | 0 | 1 |
| 1 | 1 | 1 | 1 | 1 | 1 | 1 | 1 | 1 | 1 | 0 |

Therefore,

$M_0 \equiv p \vee q \vee r,\qquad M_1 \equiv p \vee q \vee \neg r,\qquad M_2 \equiv p \vee \neg q \vee r,$

$M_3 \equiv p \vee \neg q \vee \neg r,\qquad M_4 \equiv \neg p \vee q \vee r,\qquad M_5 \equiv \neg p \vee q \vee \neg r,$

$M_6 \equiv \neg p \vee \neg q \vee r,\qquad M_7 \equiv \neg p \vee \neg q \vee \neg r.$

or, instead of the expressions above, we have

$M_{000} \equiv p \vee q \vee r,\qquad M_{001} \equiv p \vee q \vee \neg r,\qquad M_{010} \equiv p \vee \neg q \vee r,$

$M_{011} \equiv p \vee \neg q \vee \neg r,\qquad M_{100} \equiv \neg p \vee q \vee r,\qquad M_{101} \equiv \neg p \vee q \vee \neg r,$

$M_{110} \equiv \neg p \vee \neg q \vee r,\qquad M_{111} \equiv \neg p \vee \neg q \vee \neg r.$

**EXAMPLE 1.4.5** Use a truth table to find the full conjunctive normal form of $(p \wedge q) \vee (\neg p \wedge r)$.

*Solution*: Construct the truth table for $(p \wedge q) \vee (\neg p \wedge r)$ as follows.

**Table 1.4.5** Truth table for $(p \wedge q) \vee (\neg p \wedge r)$

| $p$ | $q$ | $r$ | $(p \wedge q) \vee (\neg p \wedge r)$ |
|---|---|---|---|
| 0 | 0 | 0 | 0 |
| 0 | 0 | 1 | 1 |
| 0 | 1 | 0 | 0 |
| 0 | 1 | 1 | 1 |
| 1 | 0 | 0 | 0 |
| 1 | 0 | 1 | 0 |
| 1 | 1 | 0 | 1 |
| 1 | 1 | 1 | 1 |

From Table 1.4.5, we have
$$(p \wedge q) \vee (\neg p \wedge r) \equiv (p \vee q \vee r) \wedge (p \vee \neg q \vee r) \wedge (\neg p \vee q \vee r) \wedge (\neg p \vee q \vee \neg r)$$
or
$$(p \wedge q) \vee (\neg p \wedge r) \equiv M_{000} \wedge M_{010} \wedge M_{100} \wedge M_{101}$$
or
$$(p \wedge q) \vee (\neg p \wedge r) \equiv M_0 \wedge M_2 \wedge M_4 \wedge M_5.$$

**EXAMPLE 1.4.6** Write the full conjunctive normal form for $W$ in Table 1.4.3.

*Solution*: $W \equiv (p \vee q \vee \neg r) \wedge (p \vee \neg q \vee r) \wedge (\neg p \vee q \vee \neg r) \wedge (\neg p \vee \neg q \vee \neg r)$
$\equiv M_1 \wedge M_2 \wedge M_5 \wedge M_7$

We can construct the full conjunctive normal form without using truth table, but instead, using equivalences as shown in EXAMPLE 1.4.7.

**EXAMPLE 1.4.7** Construct the full conjunctive normal form for $(p \rightarrow (q \vee r)) \wedge (p \vee q)$.

*Solution*: First, convert the wff into a conjunctive normal form by eliminating $\rightarrow$ using equivalences
$$(p \rightarrow (q \vee r)) \wedge (p \vee q) \equiv (\neg p \vee q \vee r) \wedge (p \vee q).$$
Next, add the extra variable $r$ in the second fundamental disjunction $p \vee q$.
$$(p \rightarrow (q \vee r)) \wedge (p \vee q) \equiv (\neg p \vee q \vee r) \wedge (p \vee q) \equiv (\neg p \vee q \vee r) \wedge (p \vee q \vee F)$$
$$\equiv (\neg p \vee q \vee r) \wedge (p \vee q \vee (r \wedge \neg r))$$

Then, use distributive law of $\vee$ over $\wedge$.
$$(p\rightarrow(q\vee r))\wedge(p\vee q)\equiv(\neg p\vee q\vee r)\wedge(p\vee q\vee r)\wedge(p\vee q\vee\neg r)$$

**EXAMPLE 1.4.8** Find the full conjunctive normal form for $p\vee q\rightarrow r$.

Solution: $p\vee q\rightarrow r\equiv\neg(p\vee q)\vee r\equiv(\neg p\wedge\neg q)\vee r\equiv(\neg p\vee r)\wedge(\neg q\vee r)$
$\equiv(\neg p\vee F\vee r)\wedge(F\vee\neg q\vee r)$
$\equiv(\neg p\vee(q\wedge\neg q)\vee r)\wedge((p\wedge\neg p)\vee\neg q\vee r)$
$\equiv(\neg p\vee q\vee r)\wedge(\neg p\vee\neg q\vee r)\wedge(p\vee\neg q\vee r)\wedge(\neg p\vee\neg q\vee r)$
$\equiv(\neg p\vee q\vee r)\wedge(\neg p\vee\neg q\vee r)\wedge(p\vee\neg q\vee r)$
$\equiv M_{100}\wedge M_{110}\wedge M_{010}$

**THEOREM 1.4.1** states the relationship between $m_i$ and $M_i$ with $n$ propositional variables.

**THEOREM 1.4.1** Let $m_i$ and $M_i$ be fundamental conjunction and fundamental disjunction of $n$ propositional variables, respectively. Then $\neg m_i = M_i$, $\neg M_i = m_i$.

The reader can verify this theorem using truth tables. THEOREM 1.4.1 can help us to find the full conjunctive normal form of a wff from its full disjunctive normal form, and vice versa. To see this, let
$$W\equiv m_{i_1}\vee m_{i_2}\vee\cdots\vee m_{i_k},\ 0\leqslant i_j\leqslant 2^n-1,\ j=1,2,\cdots,k,$$
and
$$\{j_1,j_2,\cdots,j_{2^n-k}\}=\{0,1,2,\cdots,2^n-1\}-\{i_1,i_2,\cdots,i_k\}.$$
Then the assignments of the corresponding binary numbers of $j_1,j_2,\cdots,j_{2^n-k}$ make $\neg W$ true. Hence
$\neg W\equiv m_{j_1}\vee m_{j_2}\vee\cdots\vee m_{j_{2^n-k}}$
$W\equiv\neg\neg W\equiv\neg(m_{j_1}\vee m_{j_2}\vee\cdots\vee m_{j_{2^n-k}})$
$\equiv\neg m_{j_1}\wedge\neg m_{j_2}\wedge\cdots\wedge\neg m_{j_{2^n-k}}$
$\equiv M_{j_1}\wedge M_{j_2}\wedge\cdots\wedge M_{j_{2^n-k}}.$

**EXAMPLE 1.4.9** Suppose that wff $A$ has three propositional variables $p,q$ and $r$. If $A$ has a full conjunctive normal form $A\equiv M_1\wedge M_2\wedge M_5\wedge M_6\wedge M_7$. Find the full disjunctive normal form of $A$.

Solution: Since $\{0,1,2,3,4,5,6,7\}-\{1,2,5,6,7\}=\{0,3,4\}$. Consequently,
$$m_0\vee m_3\vee m_4\equiv(\neg p\wedge\neg q\wedge\neg r)\vee(\neg p\wedge q\wedge r)\vee(p\wedge\neg q\wedge\neg r)$$
is the full disjunctive normal form of $A$.

**EXAMPLE 1.4.10** Suppose that wff $B$ has two propositional variables $p$, $q$, and $B \equiv m_0 \vee m_2$. Find the full conjunctive normal form of $B$.

*Solution*: $\{0,1,2,3\} - \{0,2\} = \{1,3\}$

$$B \equiv M_1 \wedge M_3 \equiv M_{01} \wedge M_{11} \equiv (p \vee \neg q) \wedge (\neg p \vee \neg q)$$

Since the full disjunctive (conjunctive) normal form of any wff $W$ is unique, we conclude that two wffs are equivalent if and only if they share the same full disjunctive (conjunctive) normal forms.

## WORDS AND EXPRESSIONS

| literal | 文字 |
| fundamental conjunction | 基本合取式 |
| fundamental disjunction | 基本析取式 |
| disjunctive normal form | 析取范式 |
| full disjunctive normal form | 主析取范式 |
| conjunctive normal form | 合取范式 |
| full conjunctive normal form | 主合取范式 |

## EXERCISES 1.4

1. Use truth table to transform each of the following wffs into the full disjunctive normal form.

   $(p \rightarrow q) \rightarrow p$   $p \rightarrow (q \rightarrow p)$   $(p \rightarrow q) \wedge r$
   $p \rightarrow q \wedge r$

2. Transform each of the following wffs into the full disjunctive normal form.

   $(\neg p \rightarrow q) \rightarrow (\neg q \vee p)$   $\neg (p \rightarrow q) \wedge q \wedge r$
   $(p \vee (q \wedge r)) \rightarrow (p \vee q \vee r)$   $p \rightarrow q \wedge r$

3. Use truth table to transform the wffs into full conjunctive normal forms.

   $\neg (p \rightarrow \neg q) \wedge \neg p$   $(p \wedge q) \vee (\neg p \vee r)$
   $(p \rightarrow (p \vee q)) \vee r$   $(p \vee q) \wedge r$

4. Use equivalences to transform each of the following wffs into the full conjunctive normal form.

   $(p \rightarrow q) \rightarrow p$   $p \rightarrow (q \rightarrow p)$   $(p \vee q) \wedge r$
   $p \rightarrow q \wedge r$   $q \wedge \neg p \rightarrow p$

5. Use full disjunctive normal form to determine whether $p \to (q \to r)$ and $q \to (p \to r)$ are equivalent.

6. Use full conjunctive normal form to determine whether $(p \to q) \wedge (p \to r)$ and $p \to (q \wedge r)$ are equivalent.

## 1.5 Functionally Complete Set of Logical Connectives

Up to now, the five connectives in the set $\{\neg, \wedge, \vee, \to, \leftrightarrow\}$ are used to form the wffs of the propositional logic. Are there any other sets of connectives that will do the same job? The answer is yes. A collection of logical connectives is called a **functionally complete set** if every wff of the propositional logic is logically equivalent to a wff involving only these connectives.

**THEOREM 1.5.1** $S = \{\neg, \wedge, \vee\}$ is a functionally complete set of logical connectives.

*Proof*: We have already seen that every wff has a unique full disjunctive normal form, which uses only the connectives $\neg, \wedge$ and $\vee$. Therefore $\{\neg, \wedge, \vee\}$ is a functionally complete set of connectives for the propositional logic.

**COROLLARY 1.5.1** The following sets are functionally complete sets of logical connectives.

1. $S_1 = \{\neg, \wedge, \vee, \to, \leftrightarrow\}$
2. $S_2 = \{\neg, \wedge, \vee, \to\}$
3. $S_3 = \{\neg, \wedge\}$
4. $S_4 = \{\neg, \vee\}$
5. $S_5 = \{\neg, \to\}$

*Proof*: $S_1$ and $S_2$ are obviously functionally complete sets, since $\{\neg, \wedge, \vee\}$ is a functionally complete set. THEOREM 1.5.1 means that we do not need implication $\to$. We need only to show that the form $A \vee B$ can be written in terms of $\neg, \wedge$. This can be seen by the equivalences

$$A \vee B \equiv \neg\neg(A \vee B) \equiv \neg(\neg A \wedge \neg B).$$

Therefore $S_3 = \{\neg, \wedge\}$ is a functionally complete set. Similarly, $S_4 = \{\neg, \vee\}$ is also a functionally complete set.

Because we have the equivalence $A \rightarrow B \equiv \neg A \lor B$, $S_5 = \{\neg, \rightarrow\}$ is a functionally complete set.

Are there any single connectives that are functionally complete sets? We do not find one among the five basic connectives. But we will introduce two new connectives which can be proved to form single functionally complete sets.

**DEFINITION 1.5.1** Let $p$ and $q$ be propositions. The proposition $p$ NAND $q$, denoted by $p \mid q$, is true when either $p$ or $q$, or both, are false; and it is false when both $p$ and $q$ are true. The connective $\mid$ is called **Sheffer stroke**. Here NAND is short for the "Negation of And".

DEFINITION 1.5.1 implies that $p \mid q \equiv \neg(p \land q)$. We can show that $\{\mid\}$ is a functionally complete set of a single connective as follows:

$\neg p \equiv \neg(p \land p) \equiv p \mid p$
$p \lor q \equiv \neg \neg (p \lor q) \equiv \neg(\neg p \land \neg q) \equiv \neg p \mid \neg q \equiv (p \mid p) \mid (q \mid q)$.

We then can conclude that $\{\mid\}$ is a functionally complete set because $\{\neg, \lor\}$ is a functionally complete set.

**DEFINITION 1.5.2** Let $p$ and $q$ be propositions. The proposition $p$ NOR $q$, denoted by $p \downarrow q$, is true when both $p$ and $q$ are false, and it is false otherwise. The connective $\downarrow$ is called **Peirce arrow**. Here NOR is short for the "Negation of OR".

From DEFINITION 1.5.2 we immediately conclude that $p \downarrow q \equiv \neg(p \lor q)$. Similarly, we can show that $\{\downarrow\}$ is a functionally complete set of a single connective in the following way:

$\neg p \equiv \neg(p \lor p) \equiv p \downarrow p$
$p \lor q \equiv \neg \neg (p \lor q) \equiv \neg(p \downarrow q) \equiv (p \downarrow q) \downarrow (p \downarrow q)$.

## WORDS AND EXPRESSIONS

| | |
|---|---|
| functionally complete set | 功能完备组 |
| NAND | 与非 |
| NOR | 或非 |

## EXERCISES 1.5

1. Represent these wffs using only NAND( $\mid$ )

$p \wedge q$ $\quad\quad p \to q$ $\quad\quad p \to (\neg p \to q)$

2. Represent these wffs involving only NOR ( $\downarrow$ )

   $p \wedge q$ $\quad\quad p \to q$ $\quad\quad p \to (\neg p \to q)$ $\quad\quad p | q$

3. Show that these equivalences are true:

   $\neg(A | B) \equiv \neg A \downarrow \neg B$ $\quad\quad \neg(A \downarrow B) \equiv \neg A | \neg B$

   where $A$ and $B$ are wffs.

4. Show that $p | q$ and $q | p$ are equivalent.

5. Show that $p | (q | r)$ and $(p | q) | r$ are not equivalent, so that the logical connective | is not associative.

## 1.6 Rules of Inference

Two important questions that arise from the study of mathematics are: (1) When is a mathematical argument correct? (2) What methods can be used to construct mathematical arguments? This section will discuss these questions.

In general, an argument starts with a list of given statements called premises, hypotheses, or antecedents, and a statement called the conclusion, or consequence of the argument. We examine these premises, say $A_1, A_2, \cdots, A_n$, and try to show that the conclusion $B$ follows logically from the given statements, that is, we try to show that if each of $A_1, A_2, \cdots, A_n$ is a true statement, then the statement $B$ is also true. To do so one way is to examine the implication

$$A_1 \wedge A_2 \wedge \cdots \wedge A_n \to B,$$

where the hypothesis is the conjunction of the $n$ premises.

The argument form is called **valid** if whenever each of the premises $A_1, A_2, \cdots, A_n$ is true, then the conclusion $B$ is also true.

Note that if any one of $A_1, A_2, \cdots, A_n$ is false, the hypothesis $A_1 \wedge A_2 \wedge \cdots \wedge A_n$ is false and the implication $A_1 \wedge A_2 \wedge \cdots \wedge A_n \to B$ is automatically true, regardless of the truth value of $B$. Consequently, one way to establish the validity of a given argument is to show that the statement $A_1 \wedge A_2 \wedge \cdots \wedge A_n \to B$ is a tautology.

When all premises used in a valid argument are true, it leads to a correct conclusion.

However, a valid argument can lead to an incorrect conclusion if one or more false premises are used within the argument.

**EXAMPLE 1.6.1** Show that the hypotheses "if Jane studies, then she will pass discrete mathematics", "if Jane dose not play tennis, then she will study" and "Jane failed discrete mathematics" lead to the conclusion "Jane played tennis".

*Solution*: let $p$ be the primitive proposition "Jane studies", $q$ the proposition "Jane plays tennis", and $r$ the proposition "Jane passes discrete mathematics". Then the premises are $A_1 = p \rightarrow r$, $A_2 = \neg q \rightarrow p$, and $A_3 = \neg r$. The conclusion is simply $B = q$.

We construct the truth table for the implication
$$(p \rightarrow r) \wedge (\neg q \rightarrow p) \wedge \neg r \rightarrow q.$$

**Table 1.6.1** Truth table for $(p \rightarrow r) \wedge (\neg q \rightarrow p) \wedge \neg r \rightarrow q$

| | | | $A_1$ | $A_2$ | $A_3$ | $A_1 \wedge A_2 \wedge A_3 \rightarrow B$ |
|---|---|---|---|---|---|---|
| $p$ | $q$ | $r$ | $p \rightarrow r$ | $\neg q \rightarrow p$ | $\neg r$ | $(p \rightarrow r) \wedge (\neg q \rightarrow p) \wedge \neg r \rightarrow q$ |
| 0 | 0 | 0 | 1 | 0 | 1 | 1 |
| 0 | 0 | 1 | 1 | 0 | 0 | 1 |
| 0 | 1 | 0 | 1 | 1 | 1 | 1 |
| 0 | 1 | 1 | 1 | 1 | 0 | 1 |
| 1 | 0 | 0 | 0 | 1 | 1 | 1 |
| 1 | 0 | 1 | 1 | 1 | 0 | 1 |
| 1 | 1 | 0 | 0 | 1 | 1 | 1 |
| 1 | 1 | 1 | 1 | 1 | 0 | 1 |

Because the final column in Table 1.6.1 contains only 1s, the implication is a tautology. Hence we can say that $A_1 \wedge A_2 \wedge A_3 \rightarrow B$ is a valid argument, or the truth of the conclusion $B$ is deduced or inferred from the truth of the premises $A_1$, $A_2$ and $A_3$.

We have seen that the truth tables are sufficient to determine whether an argument is valid. However, if an implication has three or more variables and contains several connectives, then a truth table can become quite complicated. Certainly we can use an equivalence proof, rather than a truth table, and a full disjunctive normal form to do the same job. To do this we need to introduce the basic idea of a formal reasoning system, which has three ingredients:

*A set of wffs, a set of axioms, and a set of inference rules.*

The set of wffs can be used to represent the statements of interest. For many reasoning systems to work, it needs some fundamental truths to start the process. **Axioms** (or postulates) are wffs that are the underlying assumptions about mathematical structures, which we wish to use as a basis to reason from. A **theorem** is a statement that can be shown to be true. Theorems are sometimes called **proposition facts** or **results**. Usually, axioms, the hypotheses of the theorem to be proved, and previously proved theorems form the basis from which our reasoning starts to get the conclusion of the theorem.

A reasoning system needs some rules of inference to conclude things. The rules of inference are fundamental in the development of a step by step validation of how the conclusion $B$ logically follows from the premises $A_1, A_2, \cdots, A_n$ in an implication of the form

$$A_1 \wedge A_2 \wedge \cdots \wedge A_n \to B.$$

Each rule of inference arises from a tautology of implication. In some cases, the tautology of implication is stated without proof. We will now introduce those rules that provide the justification of the steps used to validate the arguments. A common way to represent an inference rule is to draw a horizontal line, and then place the premises above the line and the conclusion below the line. The premises can be listed vertically, and the conclusion is prefixed by the symbol "$\therefore$" as follows:

$$\begin{array}{c} P_1 \\ P_2 \\ \vdots \\ P_n \\ \hline \therefore C \end{array}$$

The symbol $\therefore$ can be read as any of the following words:

therefore, thus, whence, so, ergo, hence

indicating that $C$ is the conclusion from the premises $P_1, P_2, \cdots, P_n$. For example, the tautology $A \wedge (A \to B) \to B$ is one of the basic inference rules called **modus ponens** or the **law of detachment** (*Modus Ponens* comes from Latin and may be translated as "the method of affirming"). In symbolic form this tautology is written as follows:

$$\begin{array}{c} A \\ A \to B \\ \hline \therefore B \end{array}$$

# 1 Propositional Logic

Modus Ponens states that if both an implication $A \rightarrow B$ and its hypothesis $A$ are known to be true, then the conclusion $B$ of this implication is true.

**EXAMPLE 1.6.2** Suppose that the premises "if it snows today, then we will go skiing" and "it is snowing today" are true, then by modus ponens, it follows that "we will go skiing" is true.

Table 1.6.2 lists a few more useful inference rules for the propositional calculus.

**Table 1.6.2 Useful inference rules for the propositional calculus**

| Rule of inference | Tautology | Name |
| --- | --- | --- |
| $A$<br>$A \rightarrow B$<br>$\therefore B$ | $A \wedge (A \rightarrow B) \rightarrow B$ | Modus ponens |
| $A$<br>$\therefore A \vee B$ | $A \rightarrow (A \vee B)$ | Addition |
| $A \wedge B$<br>$\therefore A$ | $A \wedge B \rightarrow A$ | Simplification |
| $A$<br>$B$<br>$\therefore A \wedge B$ | $(A) \wedge (B) \rightarrow A \wedge B$ | Conjunction |
| $A \rightarrow B$<br>$\neg B$<br>$\therefore \neg A$ | $(A \rightarrow B) \wedge \neg B \rightarrow \neg A$ | Modus tollens |
| $A \vee B$<br>$\neg A$<br>$\therefore B$ | $(A \vee B) \wedge \neg A \rightarrow B$ | Disjunctive syllogism |
| $A \rightarrow B$<br>$B \rightarrow C$<br>$\therefore A \rightarrow C$ | $(A \rightarrow B) \wedge (B \rightarrow C) \rightarrow (A \rightarrow C)$ | Hypothetical syllogism |
| $A \vee B$<br>$\neg A \vee C$<br>$\therefore B \vee C$ | $(A \vee B) \wedge (\neg A \vee C) \rightarrow (B \vee C)$ | Resolution |
| $A \rightarrow C$<br>$B \rightarrow C$<br>$\therefore (A \vee B) \rightarrow C$ | $(A \rightarrow C) \wedge (B \rightarrow C) \rightarrow (A \vee B \rightarrow C)$ | Proof by cases |

Continue

| Rule of inference | Tautology | Name |
|---|---|---|
| $A \rightarrow B$<br>$C \rightarrow D$<br>$A \lor C$<br>$\therefore B \lor D$ | $(A \rightarrow B) \land (C \rightarrow D) \land (A \lor C) \rightarrow (B \lor D)$ | Constructive dilemma |
| $A \rightarrow B$<br>$C \rightarrow D$<br>$\neg B \lor \neg D$<br>$\therefore \neg A \lor \neg C$ | $(A \rightarrow B) \land (C \rightarrow D) \land (\neg B \lor \neg D) \rightarrow (\neg A \lor \neg C)$ | Destructive dilemma |

We have introduced the three ingredients that make up any formal reasoning system. How do we reason in such a system? To describe the reasoning process in a reasonable way, we need to define the concept of proof.

A **proof** is a finite sequence of wffs with the property that each wff in the sequence either is an axiom or can be inferred from previous wffs in the sequence. The last wff in a proof is called a **theorem**. For example, suppose the following sequence of wffs is a proof:

$$W_1, W_2, \cdots, W_n,$$

where the last wff $W_n$ is a theorem. $W_1$ is an axiom because there are not any previous wffs in the sequence to infer it. $W_i (2 \leq i \leq n-1)$ is either an axiom or the conclusion of an inference rule. We demonstrate that the theorem $W_n$ is true with a sequence of statements $W_1, W_2, \cdots, W_n$ that form a proof.

We will write proofs in table format, where each line is numbered and contains a wff together with the reason why it is there. For example, a proof sequence $W_1, W_2, \cdots, W_n$ will be written as

Proof  1.  $W_1$      reason for $W_1$
       2.  $W_2$      reason for $W_2$
       $\vdots$  $\vdots$  $\vdots$
       $n$.  $W_n$   reason for $W_n$

**EXAMPLE 1.6.3** Demonstrate the validity of the argument

$$p \to r$$
$$\neg p \to q$$
$$q \to s$$
$$\therefore \neg r \to s$$

Proof:

1. $p \to r$      Premise
2. $\neg r \to \neg p$      Equivalence of step 1
3. $\neg p \to q$      Premise
4. $\neg r \to q$      Hypothetical syllogism of step 2 and step 3
5. $q \to s$      Premise
6. $\neg r \to s$      Hypothetical syllogism of step 4 and step 5

**EXAMPLE 1.6.4** Show that the premises "If you send me an e-mail message, then I will finish writing the program", "If you do not send me an e-mail message, then I will go to sleep early", and "If I go to sleep early, then I will wake up feeling refreshed" lead to the conclusion "If I do not finish writing the program, then I will wake up feeling refreshed".

Proof: Let $p, q, r, s$ denote the following primitive propositions

$p$: You send me an e-mail message

$q$: I will finish writing the program

$r$: I will go to sleep early

$s$: I will wake up feeling refreshed

Then the premises are $p \to q, \neg p \to r$, and $r \to s$. The conclusion is $\neg q \to s$. The following argument leads to the conclusion:

1. $p \to q$      Premise
2. $\neg q \to \neg p$      Contraposition of step 1
3. $\neg p \to r$      Premise
4. $\neg q \to r$      Hypothetical syllogism of step 2 and step 3
5. $r \to s$      Premise
6. $\neg q \to s$      Hypothetical syllogism of step 4 and step 5

Recall the result of $A \to (B \to C) \equiv A \wedge B \to C$ (EXERCISES 1.3 (4.b)). Let us replace $A$ by the conjunction $A_1 \wedge A_2 \wedge \cdots \wedge A_n$. Then we obtain the new result

$$A_1 \wedge A_2 \wedge \cdots \wedge A_n \to (B \to C) \equiv A_1 \wedge A_2 \wedge \cdots \wedge A_n \wedge B \to C.$$

This result tells us that to establish the validity of the argument
$$A_1 \wedge A_2 \wedge \cdots \wedge A_n \to (B \to C),$$
it is enough to establish the validity of the corresponding argument
$$A_1 \wedge A_2 \wedge \cdots \wedge A_n \wedge B \to C.$$

**EXAMPLE 1.6.5** Show that
$$(u \to r) \wedge ((r \wedge s) \to (p \vee t)) \wedge (q \to (u \wedge s)) \wedge \neg t \to (q \to p)$$
is a tautology.

*Proof*: we consider the equivalent wff:
$$(u \to r) \wedge ((r \wedge s) \to (p \vee t)) \wedge (q \to (u \wedge s)) \wedge \neg t \wedge q \to p$$

1. $q$ — Premise
2. $q \to (u \wedge s)$ — Premise
3. $u \wedge s$ — Modus ponens of step 1 and step 2
4. $u$ — Simplification of step 3
5. $u \to r$ — Premise
6. $r$ — Modus ponens of step 4 and step 5
7. $s$ — Simplification of step 3
8. $r \wedge s$ — Conjunction of step 6 and step 7
9. $(r \wedge s) \to (p \vee t)$ — Premise
10. $p \vee t$ — Modus ponens of step 8 and step 9
11. $\neg t$ — Premise
12. $p$ — Disjunctive syllogism of step 10 and step 11

Sometimes, we use the method of **proof by contradiction** to establish the validity of an argument. The idea is based on the following equivalence:
$$A_1 \wedge A_2 \wedge \cdots \wedge A_n \to B \equiv \neg(A_1 \wedge A_2 \wedge \cdots \wedge A_n) \vee B$$
$$\equiv \neg(A_1 \wedge A_2 \wedge \cdots \wedge A_n \wedge \neg B) \vee F$$
$$\equiv A_1 \wedge A_2 \wedge \cdots \wedge A_n \wedge \neg B \to F.$$

If $A_1 \wedge A_2 \wedge \cdots \wedge A_n \to B$ is a tautology, then as premises $A_1, A_2, \cdots, A_n$ and $\neg B$ lead to a false.

**EXAMPLE 1.6.6** Establish the validity of the argument
$$(\neg p \leftrightarrow q) \wedge (q \to r) \wedge \neg r \to p.$$

*Proof*: We assume the negation $\neg p$ of the conclusion $p$ as another premise. The objective now is to use these four premises to derive a contradiction $F$. Our derivation follows as:

1. $\neg p \leftrightarrow q$                 Premise
2. $(\neg p \rightarrow q) \wedge (q \rightarrow \neg p)$     Equivalence of step 1
3. $\neg p \rightarrow q$                Simplification of step 2
4. $q \rightarrow r$                  Premise
5. $\neg p \rightarrow r$                Hypothetical syllogism step 3 and step 4
6. $\neg p$                     Premise (the one assumed)
7. $r$                        Modus ponens of step 5 and step 6
8. $\neg r$                     Premise
9. false                 Conjunction of step 7 and step 8

## WORDS AND EXPRESSIONS

| | |
|---|---|
| modus ponens | 假言推理 |
| modus tollens | 拒取式 |
| hypothetical syllogism | 假言三段论 |
| resolution | 分解规则 |
| constructive dilemma | 构造性两难 |
| destructive dilemma | 破坏性两难 |

## EXERCISES 1.6

1. Use truth tables to verify that each of the following is a valid argument.

   a. $p \wedge (p \rightarrow q) \wedge r \rightarrow ((p \vee q) \rightarrow r)$
   b. $(p \wedge q \rightarrow r) \wedge \neg q \wedge (p \rightarrow \neg r) \rightarrow (\neg p \vee \neg q)$
   c. $(p \vee q \vee r) \wedge \neg q \rightarrow p \vee r$

2. Verify that each of the following is a tautology by showing that it is impossible for the conclusion to have the truth value 0 while the hypothesis has the truth value 1.

   a. $p \wedge q \rightarrow q$
   b. $p \rightarrow p \vee q$
   c. $(p \vee q) \wedge \neg p \rightarrow q$

3. Construct an argument using rules of inference to show that the hypotheses "Randy works hard", "If Randy works hard, then he is a dull boy", and "If Randy is a dull boy, then he will not get the job" imply the conclusion "Randy will not get the

job".

4. Show that the following argument is valid:
   If the band could not play rock music or the refreshments were not delivered on time, then the New Year's party would have been canceled and Alicia would have been angry. If the party was canceled, then refunds would have had to be made. No refunds were made.
   Therefore the band could play rock music.

5. Establish the validity of the following arguments.
   a. $(p \wedge \neg q) \wedge r \rightarrow (p \wedge r) \vee q$
   b. $p \wedge (p \rightarrow q) \wedge (\neg q \vee r) \rightarrow r$
   c. $(p \rightarrow q) \wedge \neg q \wedge \neg r \rightarrow \neg (p \vee r)$
   d. $(p \wedge q) \wedge (p \rightarrow (r \wedge q)) \wedge (r \rightarrow (s \vee t)) \wedge \neg s \rightarrow t$
   e. $p \wedge (p \rightarrow q) \wedge (s \vee r) \wedge (r \rightarrow \neg q) \rightarrow (s \vee t)$
   f. $(p \rightarrow q) \wedge (\neg r \vee s) \wedge (p \vee r) \rightarrow (\neg q \rightarrow s)$
   g. $((p \vee q) \rightarrow (r \wedge s)) \wedge (s \vee v \rightarrow u) \rightarrow (p \rightarrow u)$
   h. $(p \rightarrow \neg q) \wedge (\neg r \vee q) \wedge r \wedge \neg s \rightarrow \neg p$
   i. $(p \vee q) \wedge (p \rightarrow r) \wedge (q \rightarrow s) \rightarrow (r \vee s)$

# 2
# Predicate Logic

The propositional calculus provides adequate tools for reasoning about propositional wffs, which are combinations of propositions. But a proposition is a sentence taken as a whole. With this restriction, propositional logic is not powerful enough to represent all types of assertions that are used, or to express certain types of relationships between propositions such as equivalence. Consequently propositional calculus does not provide the tools to do everyday reasoning. For example, in the following argument it is impossible to find a formal way to test the correctness of the inference without further analysis of each sentence:

All men are mortal.
Socrates is a man.
Therefore, Socrates is mortal.

When dealing with such an argument, we need a new logic, first order predicate calculus, to study the inner structure of sentences. In predicate logic we break up sentences into parts and symbolize a sentence so that the information needed for reasoning is characterized in some way. A natural start point is to introduce two new features, predicates and quantifiers, to cope with the deficiencies of propositional logic.

## 2.1 Predicates and Quantifiers

The statement "$x$ is greater than $5$" has two parts: the variable $x$ is the subject of the statement, and "is greater than 5" is the **predicate** which refers to a property that the subject of the statement can have. The word "predicate" comes from the Latin word *praedicare* which means to proclaim. Let $P(x)$ represent "$x$ is greater than 5", then $P$ denotes the predicate "is greater than 5", and $x$ is the variable. The statement $P(x)$ is also said to be the value of the **propositional function** $P$ at $x$. Once a value has been assigned to the variable $x$, the statement $P(x)$ becomes a proposition and has a truth value.

**EXAMPLE 2.1.1** Let $P(x)$ denote the statement "$x$ is an even integer". What are the truth values of $P(4)$ and $P(5)$?

*Solution*: We obtain the statement $P(4)$ by setting $x=4$ in the statement $P(x)$. Thus, $P(4)$ is the statement "4 is an even integer" which is true. However, $P(5)$ is false because it is the statement "5 is an even integer".

We can also have statements that involve more than one variable. For instance, consider the statement "$x<y$". We can denote this statement by $Q(x,y)$, where $x$ and $y$ are variables and $Q$ is the predicate. When values are assigned to the variables $x$ and $y$, the statement $Q(x,y)$ has truth values.

**EXAMPLE 2.1.2** Let $Q(x,y)$ denote the statement "$x<y$". What are the truth values of the propositions $Q(3,4)$ and $Q(4,3)$?

*Solution*: To obtain $Q(3,4)$, set $x=3$ and $y=4$ in the statement "$x<y$". Hence $Q(3,4)$ is the statement "$3<4$" which is true. The statement $Q(4,3)$ is the proposition "$4<3$", which is false.

In general, a statement involving $n$ variables $x_1, x_2, \cdots, x_n$ can be denoted by
$$P(x_1, x_2, \cdots, x_n).$$
A statement of the form $P(x_1, x_2, \cdots, x_n)$ is the value of the **propositional function** $P$ at the $n$-tuple $(x_1, x_2, \cdots, x_n)$, and $P$ is called a **predicate**.

Propositional functions occur in the computer programs, as demonstrated in the following example.

**EXAMPLE 2.1.3** Consider the statement
$$\text{If } x>0 \text{ then } x:=x+3.$$
When this statement is encountered in a program, the value of the variable $x$ at the

point in the execution of the program is inserted into $P(x)$, which is "$x>0$". If $P(x)$ is true for this value of $x$, the assignment statement $x:=x+3$ is executed, so the value of $x$ is increased by 3. If $P(x)$ is false for this value of $x$, the assignment statement is not executed, so the value of $x$ remains unchanged.

When all variables in a propositional function are assigned values, the resulting statement becomes a proposition with certain truth value. However, there is another important way, called **quantification**, to create a proposition from a propositional function. Two types of quantification will be introduced; they are *universal quantification* and *existential quantification*. The area of logic that deals with predicates and quantifications is called **predicate calculus**.

Many mathematical statements assert that a property is true for all values of a variable in a particular domain, called the **universe of discourse** or **domain**. Such a statement is expressed using a universal quantification.

**DEFINITION 2.1.1** The **universal quantification** of $P(x)$ is the proposition "$P(x)$ is true for all values of $x$ in the universe of discourse".

The notation

$$\forall x\ P(x)$$

denotes the universal quantification of $P(x)$. The symbol $\forall x$ is called the **universal quantifier**. The expression $\forall x\ P(x)$ is read as

"for all $x$ $P(x)$" or "for every $x$ $P(x)$".

**EXAMPLE 2.1.4** Let $P(x)$ be the statement "$x \geq 0$". What is the truth value of the quantification $\forall x\ P(x)$, where the domain comprises all positive integers?
*Solution*: Since $P(x)$ is true for all positive integers, the quantification $\forall x\ P(x)$ is true.

**EXAMPLE 2.1.5** Let $Q(x)$ denote the statement "$x^2 - 3 > 0$". What is the truth value of the quantification $\forall x\ Q(x)$, where the universe of discourse consists of all real numbers?
*Solution*: $Q(x)$ is not true for some real number $x$, since $Q(0)$ is false. Thus $\forall x\ Q(x)$ is false.

When all elements in the universe of discourse can be listed, say, $x_1, x_2, \cdots, x_n$, the

universal quantification $\forall x\, P(x)$ is the same as the conjunction
$$P(x_1) \wedge P(x_2) \wedge \cdots \wedge P(x_n),$$
since this conjunction is true if and only if all $P(x_1), P(x_2), \cdots, P(x_n)$ are true.

**EXAMPLE 2.1.6** What is the truth value of $\forall x\, P(x)$, where $P(x)$ is the statement "$x^2 - 3x + 2 = 0$" and the domain is $\{0,1,2\}$?

*Solution*: The statement $\forall x\, P(x)$ is same as the conjunction
$$P(0) \wedge P(1) \wedge P(2).$$
Since $P(0)$ is the statement "$0^2 - 3 \times 0 + 2 = 0$", it follows that $\forall x\, P(x)$ is false.

In EXAMPLE 2.1.6 above, if the domain is $\{1,2\}$, then the truth value of $\forall x\, P(x)$ is true, because $\forall x\, P(x) \equiv P(1) \wedge P(2)$, and $P(1), P(2)$ are true. This means that specifying the universe of discourse is important when quantifiers are used. The truth value of a quantified statement depends on the elements in the domain.

**EXAMPLE 2.1.7** What is the truth value of $\forall x\, (x^2 - 3x + 2 \leq 0)$ if the universe of discourse consists of all real numbers? What is its truth value if the domain is $1 \leq x \leq 2$ instead?

*Solution*: Note that $x^2 - 3x + 2 \leq 0$ if and only if $(x-2)(x-1) \leq 0$. Therefore $x^2 - 3x + 2 \leq 0$ if and only if $1 \leq x \leq 2$. It follows that $\forall x\, (x^2 - 3x + 2 \leq 0)$ is false if the universe of discourse consists of all real numbers. However, if the domain is $\{x: 1 \leq x \leq 2\}$, $\forall x\, (x^2 - 3x + 2 \leq 0)$ is true.

To show that a statement of the form $\forall x\, P(x)$ is false, where $P(x)$ is a propositional function, we need only to find one value of $x$ in the universe of discourse for which $P(x)$ is false. Such a value of $x$ is called a counterexample to the statement $\forall x\, P(x)$.

**EXAMPLE 2.1.8** Suppose that $P(x)$ is "$x > 1$". To show that $\forall x\, P(x)$ is false, where the domain consists of all integers, we give a counterexample $x = 0$ since $0 > 1$ is not true.

Many mathematical statements assert that there is an element with certain property. Such statements are expressed using existential quantification, with which we form a proposition that is true if and only if $P(x)$ is true for at least one value of $x$ in the universe of discourse.

**DEFINITION 2.1.2** The **existential quantification** of $P(x)$ is the proposition "There exists an element $x$ in the universe of discourse such that $P(x)$ is true".

We use the notation
$$\exists x\, P(x)$$
for the existential quantification of $P(x)$. Here $\exists x$ is called the **existential quantifier**.

The existential quantification $\exists x\, P(x)$ is read as

"There is an $x$ such that $P(x)$", or

"There is at least one $x$ such that $P(x)$", or

"For some $x\, P(x)$".

**EXAMPLE 2.1.9** Let $P(x)$ denote the statement "$x^2 - 3x - 4 = 0$". The universe of discourse is $D = \{x \mid -1 < x < 4\}$. What is the truth value of $\exists x\, P(x)$?

*Solution*: Since $P(x)$ is false for every value of $D$, the existential quantification $\exists x\, P(x)$ is false.

When all elements in the universe of discourse can be listed, say, $x_1, x_2, \cdots, x_n$, the existential quantification $\exists x\, P(x)$ is the same as the disjunction $P(x_1) \vee P(x_2) \vee \cdots \vee P(x_n)$, since this disjunction is true if and only if at least one of $P(x_1), P(x_2), \cdots, P(x_n)$ is true.

Table 2.1.1 summarizes the meaning of the universal and existential quantifiers.

**Table 2.1.1  The meaning of the universal and existential quantifiers**

| Statement | When is it true? | When is it false? |
|---|---|---|
| $\forall x\, P(x)$ | For every $a$ in the domain, $P(a)$ is true | There is at least one $a$ from the universe of discourse, $P(a)$ is false |
| $\exists x\, P(x)$ | For some (at least one) $a$ in the universe of discourse, $P(a)$ is true | For every $a$ in the domain, $P(a)$ is false |

We often want to consider the negation of a quantified expression. For instance, consider the negation of the statement "Every student in the class majors in computer science". This statement is a universal quantification
$$\forall x\, P(x)$$
where $P(x)$ is the statement "$x$ majors in computer science". The negation of this statement is "it is not the case that every student in the class majors in computer science". This is simply the existential quantification of the negation of the original proposition function, namely,
$$\exists x\, \neg P(x).$$

This example illustrates the following equivalence:

$$\neg \forall x\, P(x) \equiv \exists x\, \neg P(x).$$

Suppose we wish to negate an existential quantification. For instance, consider the proposition "There is a student in this class majoring in computer science". This is the existential quantification

$$\exists x\, Q(x)$$

where $Q(x)$ denote "$x$ majors in computer science". The negation of this statement is the proposition "It is not the case that there is a student in this class majoring in computer science". This is equivalent to "every student in this class does not major in computer science", which is just the universal quantification of the negation of the original propositional function:

$$\forall x\, \neg Q(x).$$

This example illustrates the equivalence

$$\neg \exists x\, Q(x) \equiv \forall x\, \neg Q(x).$$

When the universe of discourse of a predicate $P(x)$ consists of $n$ elements, where $n$ is a positive integer, the rules for negating quantified statement are exactly the same as De Morgan's laws discussed in Section 1.3. This follows because $\neg \forall x\, P(x)$ is the same as $\neg(P(x_1) \wedge P(x_2) \wedge \cdots \wedge P(x_n))$, which is equivalent to $\neg P(x_1) \vee \neg P(x_2) \vee \cdots \vee \neg P(x_n)$ by De Morgan's laws, and this is the same as $\exists x\, \neg P(x)$. Similarly, $\neg \exists x\, P(x)$ is the same as $\neg(P(x_1) \vee P(x_2) \vee \cdots \vee P(x_n))$, which is equivalent to $\neg P(x_1) \wedge \neg P(x_2) \wedge \cdots \wedge \neg P(x_n)$, and this is the same as $\forall x\, \neg P(x)$.

**EXAMPLE 2.1.10** What are the negations of the statements "There is an honest politician" and "All people like computer science"?

*Solution*: Let $P(x)$ denote "$x$ is honest". Then the statement "There is an honest politician" is written as $\exists x\, P(x)$, where the universe of discourse consists of all politician. The negation of this statement is $\neg \exists x\, P(x)$. The negation can be translated as "every politician is dishonest". Let $Q(x)$ mean "$x$ likes computer science". Then the statement "All people like computer science" is represented by $\forall x\, Q(x)$, where the universe of discourse consists of all people. The negation can be expressed as "some person does not like computer science", and symbolized as $\exists x\, \neg Q(x)$.

**EXAMPLE 2.1.11** What are the negations of the statements $\forall x\, (x^2 - 3 \geqslant 0)$ and $\exists x$

$(x=2)$?

*Solution*: The negation of $\forall x \, (x^2 - 3 \geq 0)$ is $\neg \forall x \, (x^2 - 3 \geq 0)$, which is equivalent to $\exists x \, (x^2 - 3 < 0)$. The negation of $\exists x \, (x = 2)$ is $\neg \exists x \, (x = 2)$, which is equivalent to $\forall x \, (x \neq 2)$. The truth values of these statements depend on the universe of discourse.

Because a mathematical statement may involve more than one quantifier, we offer some examples to observe these types of statements.

**EXAMPLE 2.1.12** Assume that the universe of discourse for the variables $x$ and $y$ consists of real numbers. The commutative law for the addition of real numbers may be expressed by

$$\forall x \, \forall y \, (x + y = y + x).$$

Likewise, the statement

$$\forall x \, \exists y \, (x + y = 0)$$

says that for every real number $x$ there is a real number $y$ such that $x + y = 0$. This states that every real number has an additive inverse. Similarly, the statement

$$\forall x \, \forall y \, \forall z \, (x + (y + z) = (x + y) + z)$$

is the associative law for addition of real numbers.

**EXAMPLE 2.1.13** Let $P(x, y)$ denote the statement "There exist integers $x$ and $y$ such that $xy = 8$", where the universe of discourse is all integers. It can be represented in symbolic form by

$$\exists x \, \exists y \, (xy = 8).$$

When a statement involves both existential and universal quantifiers, however, we must be careful about the order in which the quantifiers are written.

**EXAMPLE 2.1.14** Let $Q(x, y)$ denote "$x + y = 1$", the universe of discourse is all real numbers. The statement

$$\forall x \, \exists y \, Q(x, y)$$

says that "For every real number $x$, there exists a real number $y$ such that $x + y = 1$". The statement is true. Once we have selected any $x$, the real number $y = 1 - x$ does exist and $x + y = x + (1 - x) = 1$. But we realize that each value of $x$ gives rise to a different value of $y$.

Now consider the statement

$$\exists y\ \forall x\ Q(x,y).$$

This statement is read "There exists a real number $y$ so that for all real numbers $x$, $x + y = 1$". This statement is false, since once a real number $y$ is selected, the *only* value that $x$ can have is $1 - y$.

EXAMPLE 2.1.14 illustrates that the order in which quantifiers appear makes differences. The statements $\forall x\ \exists y\ Q(x,y)$ and $\exists y\ \forall x\ Q(x,y)$ are not logically equivalent. The statement $\forall x\ \exists y\ Q(x,y)$ is true if and only if for every value of $x$ there is a value of $y$ for which $Q(x,y)$ is true. So, for this statement to be true, no matter which $x$ you choose, there must be a value of $y$ (possibly depending on the $x$ you chosen). On the other hand, $\exists y\ \forall x\ Q(x,y)$ is true if and only if there is $y$ that makes $Q(x,y)$ true for every $x$. So, for this statement to be true there must be a particular value of $y$ for which $Q(x,y)$ is true regardless of the choice of $x$.

Statements involving nested quantifiers can be negated by successively applying the rules for negating statements involving a single quantifier.

EXAMPLE 2.1.15 What is the negation of the statement $\forall x\ \exists y\ (xy = 1)$ so that no negation precedes a quantifier?

*Solution*: By successively applying the rules for negating quantified statements, we can move the negations inside all quantifiers in $\neg \forall x\ \exists y\ (xy = 1)$. In fact $\neg \forall x\ \exists y\ (xy = 1)$ is equivalent to $\exists x\ \neg \exists y\ (xy = 1)$, which is in turn equivalent to $\exists x\ \forall y\ \neg(xy = 1)$. Since $\neg(xy = 1)$ can be written in a simpler form $(xy \neq 1)$, we conclude that our negated statement can be expressed as $\exists x\ \forall y\ (xy \neq 1)$.

EXAMPLE 2.1.16 Express the negation of the statement $\forall x\ \exists y\ (P(x,y) \wedge Q(x,y) \rightarrow R(x,y))$ so that no negation precedes a quantifier.

*Solution*:
$\neg(\forall x\ \exists y\ (P(x,y) \wedge Q(x,y) \rightarrow R(x,y)))$
$\equiv \exists x\ (\neg \exists y\ (P(x,y) \wedge Q(x,y) \rightarrow R(x,y)))$
$\equiv \exists x\ \forall y\ \neg(P(x,y) \wedge Q(x,y) \rightarrow R(x,y))$
$\equiv \exists x\ \forall y\ \neg(\neg(P(x,y) \wedge Q(x,y)) \vee R(x,y))$
$\equiv \exists x\ \forall y\ (P(x,y) \wedge Q(x,y) \wedge \neg R(x,y))$

## WORDS AND EXPRESSIONS

| | |
|---|---|
| predicate | 谓词 |
| quantifier | 量词 |
| propositional function | 命题函数 |
| universe of discourse | 论域 |
| universal quantifier | 全称量词 |
| counterexample | 反例 |
| existential quantifier | 存在量词 |

## EXERCISES 2.1

1. Let $P(x)$, $Q(x)$ and $R(x)$ denote the following propositional functions:
$$P(x): x \leqslant 3, \quad Q(x): x+1 \text{ is odd}, \quad R(x): x>0.$$
   If the universe of discourse consists of all integers, what are the truth values of the following statements?
   a. $\neg(P(-4) \lor Q(-3))$
   b. $P(3) \lor Q(3) \lor \neg R(3)$
   c. $P(2) \land Q(2) \to R(2)$
   d. $P(0) \to (\neg Q(-1) \leftrightarrow R(1))$

2. Translate the statements into English, where $C(x)$ is "$x$ is a comedian" and $F(x)$ is "$x$ is funny" and the universe of discourse consists of all people.
   a. $\forall x \, (C(x) \to F(x))$
   b. $\forall x \, (C(x) \land F(x))$
   c. $\exists x \, (C(x) \to F(x))$
   d. $\exists x \, (C(x) \land F(x))$

3. Let $P(x)$ be the statement "$x = x^2$". If the universe of discourse consists of the integers, what are the truth values of the following statements?
   a. $P(0)$
   b. $P(1)$
   c. $\forall x \, \neg P(x)$
   d. $\exists x \, P(x)$

e. $\forall x\, P(x)$
f. $\exists x\, \neg P(x)$

4. Determine the truth value of each of these statements if the universe of discourses is all real numbers.
   a. $\exists x\, (x^2 = 2)$
   b. $\exists x\, (x^2 = -1)$
   c. $\forall x\, (x^2 + 2 \geq 1)$
   d. $\forall x\, (x^2 \neq x)$

5. Suppose that the universe of discourse of $P(x)$ consists of integers 0, 1, 2, 3, and 4. Write each of the following propositions using disjunctions, conjunctions and negations.
   a. $\exists x\, P(x)$
   b. $\forall x\, P(x)$
   c. $\exists x\, \neg P(x)$
   d. $\forall x\, \neg P(x)$
   e. $\neg \exists x\, P(x)$
   f. $\neg \forall x\, P(x)$

6. Let $P(x,y)$ and $Q(x,y)$ denote the following propositional functions
   $$P(x,y): x^2 \geq y,\ Q(x,y): x + 2 < y.$$
   If the universes of discourse for $x$ and $y$ are the set of all real numbers, what are the truth values of the following statements?
   a. $P(2,2) \to Q(1,1)$
   b. $\exists y\, Q(1,y)$
   c. $\exists x\, P(x,2)$
   d. $\exists x\, \exists y\, P(x,y)$
   e. $\forall x\, \exists y\, Q(x,y)$
   f. $\exists y\, \forall x\, Q(x,y)$
   g. $\forall x\, \forall y\, P(x,y)$
   h. $\exists x\, \exists y\, (P(x,y) \wedge Q(x,y))$

7. Suppose that the variables of $P(x,y)$ can take values in the domain $D = \{0,1,2\}$. Write each of these propositions using disjunctions, conjunctions and negations.
   a. $\forall x\, \forall y\, P(x,y)$

b. $\exists x \, \exists y \, P(x,y)$
c. $\forall x \, \exists y \, P(x,y)$
d. $\exists x \, \forall y \, P(x,y)$
e. $\exists y \, \forall x \, P(x,y)$

8. Write a quantified expression over some domain to denote each of the following propositions.
   a. $Q(0) \land Q(1)$
   b. $Q(2) \land Q(4) \land Q(6) \land Q(8) \land \cdots$
   c. $Q(1) \lor Q(3) \lor Q(5) \lor Q(7) \lor \cdots$
   d. $P(2,2) \lor P(2,3) \lor P(2,5)$
   e. $(P(2,2) \land P(2,3) \land P(2,5)) \lor (P(3,2) \land P(3,3) \land P(3,5)) \lor (P(5,2) \land P(5,3) \land P(5,5))$

9. Negate and simplify each of the following.
   a. $\exists x \, (P(x) \lor Q(x))$
   b. $\forall x \, (P(x) \land \neg Q(x))$
   c. $\forall x \, (P(x) \rightarrow Q(x))$
   d. $\exists x \, (P(x) \lor Q(x) \rightarrow R(x))$
   e. $\forall x \, \exists y \, R(x,y)$
   f. $\forall y \, \forall x \, (C(x,y) \lor F(x,y))$
   g. $\forall x \, \exists y \, \forall z \, H(x,y,z) \land \exists z \, \forall y \, H(x,y,z)$
   h. $\exists x \, \exists y \, \neg R(x,y) \land \forall x \, \forall y \, Q(x,y)$
   i. $\exists x \, \exists y \, (H(x,y) \leftrightarrow F(x,y))$

## 2.2 Well-Formed Formulas in Predicate Logic

To give a precise description of predicate logic, we need an alphabet of symbols. We will use several kinds of letters and symbols, described as follows:

Individual variables: $x, y, z, \cdots, x_i, y_i, z_i, \cdots, i \geq 1$
Individual constants: $a, b, c, \cdots, a_i, b_i, c_i, \cdots, i \geq 1$
Functional constants: $f, g, h, \cdots, f_i, g_i, h_i, \cdots, i \geq 1$
Predicate constants: $P, Q, R, \cdots, P_i, Q_i, R_i, \cdots, i \geq 1$

Connective symbols: $\neg, \wedge, \vee, \rightarrow, \leftrightarrow$

Quantifier symbols: $\forall, \exists$

Punctuation symbols: ( , )

From time to time, we will use other letters, or strings of letters to denote variables or constants. We will also allow letters to be subscripted. The number of arguments for a predicate or function will normally be clear from the text. A predicate with no arguments is considered to be a proposition.

**DEFINITION 2.2.1** A **term** is either a variable, a constants, or a function applied to arguments that are terms.

For example, $x$, $a$ and $f(x, g(b))$ are terms.

**DEFINITION 2.2.2** An **atomic formula** (or simply **atom**) is a predicate applied to arguments that are terms.

For example, $P(x)$, $Q(x,a)$, and $R(y,z,f(a))$ are atoms.

We can define the wff (the well-formed formula) of the predicate calculus inductively as follows.

**DEFINITION 2.2.3**
1. Any atom is a wff.
2. If $A$ is a wff, then $\neg A$ is also a wff.
3. If $A$ and $B$ are wffs, then $A \wedge B, A \vee B, A \rightarrow B, A \leftrightarrow B$ are also wffs.
4. If $A$ is a wff, and $x$ is a variable, then $\exists x\, A$ and $\forall x\, A$ are wffs.

For example $P(x)$, $\exists x\, P(x,y)$, $\forall x\, P(x,y,z) \rightarrow Q(x)$, $\neg \forall x\, R(x)$ are wffs. To write formulas without too many parentheses, we will agree that the quantifiers have the same precedence as the negation symbol. We will continue to use the same hierarchy of the precedence for the operators $\neg, \wedge, \vee, \rightarrow$ and $\leftrightarrow$. For example, $\exists x\, \neg P(x,y) \rightarrow Q(x) \vee R(y)$ means $\exists x\, (\neg P(x,y)) \rightarrow (Q(x) \vee R(y))$.

Now let us discuss the relationship between the quantifiers and the variables that appear in a wff. When a quantifier occurs in a wff, it influences some occurrences of the quantified variables. The extent of this influence is called the *scope* of the quantifier, which we formally define as follows.

**DEFINITION 2.2.4** In the wff $\forall x\, W$, $W$ is the **scope** of the quantifier $\forall x$. In the wff $\exists x\, W$, $W$ is the **scope** of the quantifier $\exists x$. An occurrence of the variable $x$ in a wff is said to be **bound** if it lies in the scope $W$ of either $\forall x$ or $\exists x$ or if it is the quantifier variable $x$ itself or when we assign a value to this variable. Otherwise, an occurrence of $x$ is said to be **free** in the wff.

**EXAMPLE 2.2.1** Consider the wff:
$$\exists x\, P(x,y) \to Q(x).$$
The scope of $\exists x$ is $P(x,y)$. The first two occurrences of $x$ are bound. The only occurrence of $y$ is free, and the third occurrence of $x$ is free.

**EXAMPLE 2.2.2** For each of the following wffs, label each occurrence of the variables as either bound or free, and find out the scope of each quantifier.
1. $\forall x\, (P(x,y) \to Q(x,z))$
2. $\forall x\, (P(x) \to Q(y)) \to \exists y\, (R(x) \wedge H(x,y,z))$

*Solution*:
1. The scope of $\forall x$ is $P(x,y) \to Q(x,z)$. In this wff, the three occurrences of $x$ are bound. The only occurrences of $y$ and $z$ are free.
2. The scope of $\forall x$ is $P(x) \to Q(y)$, the scope of $\exists y$ is $R(x) \wedge H(x,y,z)$. In this wff, the first two occurrences of $x$ are bound; the third and the fourth occurrences of $x$ are free. The first occurrence of $y$ is free, but the second and the third occurrences of $y$ are bound. The only occurrence of $z$ is free.

Translating sentences in English into logical expressions is a crucial task in mathematics, logic programming, artificial intelligence, software engineering, and many other disciplines. In propositional logic, when we use propositions to express sentences, we purposely avoid sentences whose translations require predicates and quantifiers. Translating from English to logical expression becomes even more complex when quantifiers are needed. Also, there can be many ways to translate a particular sentence. We will use some examples to illustrate how to translate sentences from English into logical expressions.

**EXAMPLE 2.2.3** Express the statement "Every student in this class has studied calculus" using predicates and quantifiers.
*Solution*: First, we rewrite the statement to clearly identify the appropriate quantifiers to

be used

"For every student in this class, that student has studied calculus".
Next, we introduce a variable $x$ so that our statement becomes

"For every student $x$ in this class, that $x$ has studied calculus".
We then introduce the predicate $P(x)$ as the statement "$x$ has studied calculus". Consequently, if the universe of discourse for $x$ consists of the students in the class, we can translate our statement as $\forall x\ P(x)$.

However, there are other correct approaches using different universe of discourse and other predicates. The approach we select depends on the subsequent reasoning we want to carry out. For example, we may be interested in a wider group of people than only those in this class. If we change the universe of discourse to all people, we will need to express our statement as:

"For every person $x$, if person $x$ is a student in this class, then $x$ has studied calculus".
If we introduce a special predicate $C(x)$ to represent "person $x$ is in this class", then our statement can be expressed as $\forall x\ (C(x) \rightarrow P(x))$.

Note that our statement can not be expressed as $\forall x\ (C(x) \wedge P(x))$ since this statement says that all people are in this class and have studied calculus. When we are interested in the background of people in subjects besides calculus, we may prefer to use a two-variable predicate $Q(x,y)$ for statement "student $x$ has studied subject $y$" and use an individual constant $a$ representing "calculus". Then we would obtain

$$\forall x\ (C(x) \rightarrow Q(x,a))\ \text{or}\ \forall x\ (C(x) \rightarrow Q(x, \text{calculus})).$$

**EXAMPLE 2.2.4** Express the statement "some student in this class has visited Beijing".

*Solution*: The statement "some student in this class has visited Beijing" means that "There is a student in this class with the property that the student has visited Beijing". By introducing a variable $x$ our statement becomes

"There is a student $x$ in this class, $x$ has visited Beijing".
We introduce the predicate $R(x)$ to represent the statement "$x$ has visited Beijing". If the universe of discourse for $x$ consists of the students in this class, we can translate this statement as $\exists x\ R(x)$.

However, if we are interested in all people not just those in this class, we will look at the statement in a slightly different way. Our statement can instead be expressed as

"There is a person $x$, $x$ is a student in this class and $x$ has visited Beijing".

Now, the universe of discourse for the variable $x$ consists of all people. We introduce the predicate $S(x)$ to represent "$x$ is a student in this class". Our solution becomes $\exists x\,(S(x) \wedge R(x))$.

Note that our statement cannot be expressed as $\exists x\,(S(x) \rightarrow R(x))$ which is true when there is someone not in the class.

EXAMPLE 2.2.3 and EXAMPLE 2.2.4 showed how quantifiers can be used to translate sentences into logical expressions. We now address the sentences that require nested quantifiers to be translated into logical expressions.

**EXAMPLE 2.2.5** Express the statement "if a person is female and is a parent, then this person is someone's mother" as a logical expression involving predicates, quantifiers with a universe of discourse consisting of all people and logical connectives.

*Solution*: The statement can be expressed as "For every person $x$, if $x$ is female and $x$ is a parent, then there exists a person $y$ such that $x$ is the mother of $y$". We use the predicates $F(x)$ to represent "$x$ is female", $P(x)$ to represent "$x$ is a parent", and $M(x,y)$ to represent "$x$ is the mother of $y$". The original statement can be represented as
$$\forall x\,(F(x) \wedge P(x) \rightarrow \exists y\,M(x,y)).$$

**EXAMPLE 2.2.6** Express the statement "Everyone has exactly one best friend" as a logical expression involving predicates, quantifiers with universe of discourse consisting of all people, and logical connectives.

*Solution*: This statement can be expressed as "For every person $x$, $x$ has exactly one best friend". To say that $x$ has exactly one best friend means that there is a person $y$ who is the best friend of $x$, and that for every person $z$, if person $z$ is not person $y$, then $z$ is not the best friend of $x$. We use the predicate $B(x,y)$ to denote the statement "$y$ is the best friend of $x$". The statement that $x$ has exactly one best friend can be represented as
$$\exists y\,(B(x,y) \wedge \forall z\,((y \neq z) \rightarrow \neg B(x,z))).$$

Consequently, our original statement can be expressed as
$$\forall x\,\exists y\,(B(x,y) \wedge \forall z\,((y \neq z) \rightarrow \neg B(x,z))).$$

**EXAMPLE 2.2.7**  Translate the statement "Every real number except zero has a multiplicative inverse".

*Solution*: We rewrite this statement as "For every real number $x$, if $x \neq 0$, then there exists a real number $y$ such that $xy = 1$". This can be rewritten as
$$\forall x \ (x \neq 0 \rightarrow \exists y \ (xy = 1)).$$

**EXAMPLE 2.2.8**  Express the definition of a limit using quantifiers.

*Solution*: Recall that the definition of the limit
$$\lim_{x \to a} F(x) = L$$
is: "For every real number $\varepsilon > 0$, there exists a real number $\delta > 0$ such that $|F(x) - L| < \varepsilon$ whenever $0 < |x - a| < \delta$". This definition can be phrased in terms of quantifiers by
$$\forall \varepsilon \ \exists \delta \ \forall x \ (0 < |x-a| < \delta \rightarrow |F(x) - L| < \varepsilon)$$
where the universes of discourse for the variables $\varepsilon$ and $\delta$ consist of all positive real numbers and the universe of discourse for $x$ consists of all real numbers.

This definition can also be expressed as
$$\forall \varepsilon > 0 \ \exists \delta > 0 \ \forall x \ (0 < |x-a| < \delta \rightarrow |F(x) - L| < \varepsilon)$$
where the universes of discourse for the variables $\varepsilon, \delta$ and $x$ consist of all real numbers.

## WORDS AND EXPRESSIONS

| | |
|---|---|
| individual variable | 个体变元 |
| individual constant | 个体常元 |
| function constant | 函数常元 |
| predicate constant | 谓词常元 |
| term | 项 |
| atom formula | 原子公式 |
| scope | 辖域 |
| bound | 约束的 |
| free | 自由的 |

## EXERCISES 2.2

1. Determine whether the following expressions are wffs.

a. $\exists x\, P(x) \rightarrow \forall x\, Q(x,y)$
b. $\exists x\, \forall y\, (P(y) \rightarrow Q(f(x),y,z))$

2. For each of the following wffs, label each occurrence of the variables as either bound or free, and find the scopes of the quantifiers.
   a. $P(x,y) \vee (\forall y\, Q(y) \rightarrow \exists x\, R(x,y))$
   b. $\forall y\, Q(y) \vee \neg P(x,y)$
   c. $\neg Q(x,y) \vee \exists x\, P(x,y)$
   d. $\forall x\, \forall y\, (P(x,y) \wedge Q(y,z)) \wedge \exists x\, P(x,y)$

3. For the universe of discourse of all integers, let $P(x), Q(x), R(x), S(x)$ and $T(x)$ be the following propositional functions.
   $P(x): x > 0$;
   $Q(x): x$ is even;
   $R(x): x$ is a perfect square;
   $S(x): x$ is divisible by 4;
   $T(x): x$ is divisible by 5.
   a. Translate the following statements into logical expressions.
      i. At least one integer is even.
      ii. There exists a positive integer that is even.
      iii. If $x$ is even, then $x$ is not divisible by 5.
      iv. No even integer is divisible by 5.
      v. There exists an even integer divisible by 5.
      vi. If $x$ is even and $x$ is a perfect square, then $x$ is divisible by 4.
   b. Determine whether each of the six statements in part (a) is true or false. For each false statement, provide a counter example.
   c. Express each of the following statement by a simple English sentence.
      i. $\forall x\, (R(x) \rightarrow P(x))$
      ii. $\exists x\, (S(x) \wedge \neg R(x))$
      iii. $\forall x\, (\neg R(x) \vee \neg Q(x) \vee S(x))$
      iv. $\forall x\, (S(x) \rightarrow \neg T(x))$

4. Let $P(x,y)$ be the statement "$x$ likes $y$" where the universe of discourse for both $x$ and $y$ consists of all people. Use quantifiers to express each of these statements:

a. Everybody likes Ann.
b. Everybody likes somebody.
c. There is somebody whom everybody likes.
d. Nobody likes everybody.
e. There is somebody whom Jack does not like.
f. There is somebody whom no one likes.
g. There is exactly one person whom everybody likes.

5. Write down an expression for each of the following statements as a quantified wff.

    a. Every natural number other than 0 has a predecessor.
    b. Any two nonzero natural numbers have a common divisor.

6. Express each of these statements using mathematical and logical operators, predicates and quantifiers, where the universe of discourse consists of all integers.

    a. The sum of two negative integers is negative.
    b. The difference of two positive integers is not necessarily positive.
    c. The sum of the squares of two integers is greater than or equal to the square of the sum.
    d. The absolute value of the product of two integers is the product of their absolute values.

7. Express $\lim_{n \to \infty} a_n = a$ in logical statement.

## 2.3  Equivalent Formulas

For a wff to have a meaning, we must give an interpretation to its symbols so that the wff can be interpreted as a statement that is true or false. For example, let's give an interpretation to the wff

$$\forall x \, \exists y \, P(x,y).$$

Let $P(x,y)$ denote that the successor of $x$ is $y$, where the variables $x$ and $y$ take values from the set of natural numbers. With this interpretation the wff becomes the statement "For every natural number $x$ there exists a natural number $y$ such that the successor of $x$ is $y$", which is true.

We can give another interpretation to the wff. Let $P(x,y)$ mean that the successor of $y$

is $x$, where $x$ and $y$ take values from all natural numbers. Then the wff means "For every natural number $x$ there exists a natural number $y$ such that the successor of $y$ is $x$", which is false.

**DEFINITION 2.3.1** An **interpretation** $I$ for a wff consists of a nonempty set $D$, called the domain of the interpretation, together with an assignment that associates the symbols of the wff to values in $D$ as follows:

1. Each predicate letter must be assigned some relation over $D$. A predicate with no arguments is a proposition and must be assigned a truth value.
2. Each function letter must be assigned a function over $D$.
3. Each free variable must be assigned a value in $D$. All free occurrences of a variable $x$ are assigned the same value in $D$.
4. Each constant must be assigned a value in $D$. All occurrences of the same constant are assigned the same value in $D$.

**EXAMPLE 2.3.1** Let $W = \forall x \, \forall y \, (F(f(x,a),y) \rightarrow F(g(y,a),x))$. One interpretation for the wff can be made as follows: Suppose $D = \{0,1,2,\cdots\}$ is the domain, $F(x,y)$ is "$x = y$", the individual constant $a = 0$, and $f, g$ are functions defined by $f(x,y) = x+y$, $g(x,y) = xy$. With this interpretation, $W$ can be written as:
$$\forall x \, \forall y \, ((x+0 = y) \rightarrow (y \cdot 0 = x)).$$
A bit of checking will convince us that $W$ is false with respect to this interpretation.

**EXAMPLE 2.3.2** Let $W = \forall x \, \forall y \, (F(f(x,a),y) \rightarrow F(g(y,a),x))$. Set the domain to be $D = \{0\}$, and $f(x,y) = x+y$, $g(x,y) = xy$, let $F(x,y)$ mean $x = y$ and $a = 0$. With this interpretation, $W$ can be expressed as
$$\forall x \, \forall y \, ((x+0 = y) \rightarrow (y \cdot 0 = x))$$
$$\equiv (0+0 = 0) \rightarrow (0 \cdot 0 = 0)$$
which is true.

**EXAMPLE 2.3.3** Given an interpretation as follows: Let domain $D = \{2,3\}$, individual constant $a = 2$ and $F(2) = 0, F(3) = 1, G(2,2) = G(3,2) = 1, L(2,2) = L(3,3) = 1, L(2,3) = 0, L(3,2) = 0$. Determine the truth values of the following statements.

1. $\forall x \, (F(x) \wedge G(x,a))$
2. $\forall x \, \exists y \, L(x,y)$

3. $\exists y \, \forall x \, L(x,y)$

*Solution*:

1. $\quad \forall x \, (F(x) \wedge G(x,a))$
   $\equiv (F(2) \wedge G(2,2)) \wedge (F(3) \wedge G(3,2))$
   $\equiv (0 \wedge 1) \wedge (1 \wedge 1)$
   $\equiv 0$

2. $\quad \forall x \, \exists y \, L(x,y)$
   $\equiv \forall x \, (L(x,2) \vee L(x,3))$
   $\equiv (L(2,2) \vee L(2,3)) \wedge (L(3,2) \vee L(3,3))$
   $\equiv (1 \vee 0) \wedge (0 \vee 1)$
   $\equiv 1$

3. $\quad \exists y \, \forall x \, L(x,y)$
   $\equiv \exists y \, (L(2,y) \wedge L(3,y))$
   $\equiv (L(2,2) \wedge L(3,2)) \vee (L(2,3) \wedge L(3,3))$
   $\equiv (1 \wedge 0) \vee (0 \wedge 1)$
   $\equiv 0$

**DEFINITION 2.3.2** A wff is **valid** if it's true for all possible interpretations. Otherwise, the wff is **invalid**.

**DEFINITION 2.3.3** A wff is **unsatisfiable** if it's false for all possible interpretations. Otherwise, it is **satisfiable**.

**EXAMPLE 2.3.4** Show that the wff $W = \exists x \, \forall y \, (P(x) \rightarrow Q(x,y))$ is satisfiable and invalid.

*Proof*: To see the wff is satisfiable, notice that the wff is true with respect to the following interpretations: $D_1 = \{2,3\}, P(2) = P(3) = 1, Q(2,2) = Q(2,3) = 1, Q(3,2) = Q(3,3) = 0$,

$W = \exists x \, \forall y \, (P(x) \rightarrow Q(x,y))$
$\equiv \exists x \, ((P(x) \rightarrow Q(x,2)) \wedge (P(x) \rightarrow Q(x,3)))$
$\equiv ((P(2) \rightarrow Q(2,2)) \wedge (P(2) \rightarrow Q(2,3))) \vee ((P(3) \rightarrow Q(3,2)) \wedge (P(3) \rightarrow Q(3,3)))$
$\equiv (1 \wedge 1) \vee (0 \wedge 0)$
$\equiv 1$.

Thus, $\exists x\, \forall y\, (P(y) \to Q(x,y))$ is satisfiable. But with the interpretation $D_2 = \{2\}$, $P(2) = 1$, $Q(2,2) = 0$,
$$W = \exists x\, \forall y\, (P(x) \to Q(x,y)) \equiv P(2) \to Q(2,2) \equiv 0.$$
In this case, the wff is false. Therefore, the wff $W$ is invalid.

**EXAMPLE 2.3.5** Show that $\forall x\, A(x) \to \exists x\, A(x)$ is valid.

*Proof*: Let $I$ be an arbitrary interpretation and $D$ the domain. If there is an element $a \in D$ such that $A(a)$ is false, then $\forall x\, A(x)$ is false. In this case $\forall x\, A(x) \to \exists x\, A(x)$ is true. If for every $a \in D$, $A(a)$ is true, then $\forall x\, A(x)$ and $\exists x\, A(x)$ are true. Thus, $\forall x\, A(x) \to \exists x\, A(x)$ is true. Therefore, $\forall x\, A(x) \to \exists x\, A(x)$ is valid.

**DEFINITION 2.3.4** Two wffs $A$ and $B$ are **equivalent** if they both have the same truth value with respect to every interpretation of both $A$ and $B$.

By an interpretation of both $A$ and $B$, we mean that all free variables, constants, functions and predicates that occur in either $A$ or $B$ are interpreted with respect to a single domain. We denote the fact that $A$ and $B$ are equivalent by $A \equiv B$.

Now, we'll give some kinds of equivalences.

1. Two equivalences relating quantifiers and negation.

    For any wff $W$ we have two equivalences, which have been discussed in Section 2.1.
    $$\neg(\forall x\, W) \equiv \exists x\, \neg W, \quad \neg(\exists x\, W) \equiv \forall x\, \neg W$$

2. Generalization of propositional wff.

    A wff $W$ is an **instance** of a propositional wff $V$ if $W$ is obtained from $V$ by replacing each propositional variable of $V$ by a wff, where all occurrences of each propositional variable in $V$ are replaced by the same wff. For example, the wff $\exists x\, P(x) \to \exists x\, P(x) \vee Q(x)$ is an instance of $P \to P \vee Q$ by replacing $P$ by $\exists x\, P(x)$ and replacing $Q$ by $Q(x)$.

    Two wffs are equivalent if they are instances of two equivalent propositional wffs, where both instances are obtained by using the same replacement of propositional letters. For example,
    $$\forall x\, (P(x) \to Q(x)) \equiv \forall x\, (\neg P(x) \vee Q(x));$$
    $$\exists x\, H(x) \wedge \neg(\exists x\, H(x)) \equiv F.$$

3. In Section 2.1, we have the conclusion that, if all elements in the domain can be

listed, namely, $D = \{a_1, a_2, \cdots, a_n\}$, then
$$\forall x\, A(x) \equiv A(a_1) \wedge A(a_2) \wedge \cdots \wedge A(a_n);$$
$$\exists x\, A(x) \equiv A(a_1) \vee A(a_2) \vee \cdots \vee A(a_n).$$

4. Equivalences with Restrictions.

If $x$ does not occur in the wff $B$, then the following equivalences hold:
$$\forall x\, (A(x) \vee B) \equiv \forall x\, A(x) \vee B;$$
$$\forall x\, (A(x) \wedge B) \equiv \forall x\, A(x) \wedge B;$$
$$\forall x\, (A(x) \to B) \equiv \exists x\, A(x) \to B;$$
$$\forall x\, (B \to A(x)) \equiv B \to \forall x\, A(x);$$
$$\exists x\, (A(x) \vee B) \equiv \exists x\, A(x) \vee B;$$
$$\exists x\, (A(x) \wedge B) \equiv \exists x\, A(x) \wedge B;$$
$$\exists x\, (A(x) \to B) \equiv \forall x\, A(x) \to B;$$
$$\exists x\, (B \to A(x)) \equiv B \to \exists x\, A(x).$$

Here, some equivalences can be easily derived from other equivalences. For example,
$$\forall x\, (A(x) \to B) \equiv \forall x\, (\neg A(x) \vee B) \equiv \forall x\, \neg A(x) \vee B$$
$$\equiv \neg \exists x\, A(x) \vee B \equiv \exists x\, A(x) \to B.$$

5. Equivalences for Quantified statement
$$\forall x\, (A(x) \wedge B(x)) \equiv \forall x\, A(x) \wedge \forall x\, B(x),$$
$$\exists x\, (A(x) \vee B(x)) \equiv \exists x\, A(x) \vee \exists x\, B(x).$$

6. We have two equivalences involving more than one quantifiers
$$\forall x\, \forall y\, W \equiv \forall y\, \forall x\, W,\ \exists x\, \exists y\, W \equiv \exists y\, \exists x\, W.$$

The following two rules can help us to obtain many equivalences.

**Replacement rule**: If $\Phi(A)$ is a wff with $A$ as its subwff, and $A \equiv B$, then $\Phi(A) \equiv \Phi(B)$, where $A$ and $B$ are wffs in predicate logic.

For example,
$$(\forall x\, P(x) \to \exists y\, Q(y)) \vee R(x,y) \equiv (\neg \forall x\, P(x) \vee \exists y\, Q(y)) \vee R(x,y)$$
because $A = \forall x\, P(x) \to \exists y\, Q(y)$ is equivalent to $B = \neg \forall x\, P(x) \vee \exists y\, Q(y)$.

**Renaming rule**: If $y$ does not occur in $A(x)$, then the following equivalences hold:
$$\exists x\, A(x) \equiv \exists y\, A(y);$$
$$\forall x\, A(x) \equiv \forall y\, A(y).$$

Remember that $A(y)$ is obtained from $A(x)$ by replacing all free occurrences of $x$ by $y$. For example, using renaming rule, we have the following expressions:

$$\forall x\ \exists y\ (P(x,y) \to \exists x\ Q(x,y) \lor \forall y\ R(x,y))$$
$$\equiv \forall z\ \exists y\ (P(z,y) \to \exists x\ Q(x,y) \lor \forall y\ R(z,y))$$
$$\equiv \forall z\ \exists u\ (P(z,u) \to \exists x\ Q(x,u) \lor \forall y\ R(z,y)).$$

In the first expression, the first $y$ in $\exists y$, the second $y$ in $P(x,y)$ and the third $y$ in $Q(x,y)$ are bound by the existential quantifier. But in last expression, the three $y$ are renamed to $u$. Similarly, in the original expression, the occurrences of $x$ except two $x$ in $\exists x\ Q(x,y)$ are renamed as $z$.

**EXAMPLE 2.3.6** Show the following equivalences.
1. $\exists x\ (P(x) \to Q(x)) \equiv \forall x\ P(x) \to \exists x\ Q(x)$
2. $\neg \exists x\ (P(x) \land Q(x)) \equiv \forall x\ (P(x) \to \neg Q(x))$
3. $\neg \forall x\ \forall y\ (P(x) \land Q(y) \to R(x,y)) \equiv \exists x\ \exists y\ (P(x) \land Q(y) \land \neg R(x,y))$

*Proof*:
1. $\exists x\ (P(x) \to Q(x)) \equiv \exists x\ (\neg P(x) \lor Q(x)) \equiv \exists x\ \neg P(x) \lor \exists x\ Q(x)$
$\equiv \neg \forall x\ P(x) \lor \exists x\ Q(x) \equiv \forall x\ P(x) \to \exists x\ Q(x)$
2. $\neg \exists x\ (P(x) \land Q(x)) \equiv \forall x\ (\neg P(x) \lor \neg Q(x)) \equiv \forall x\ (P(x) \to \neg Q(x))$
3. $\quad \neg \forall x\ \forall y\ (P(x) \land Q(y) \to R(x,y))$
$\equiv \exists x\ \neg \forall y\ (P(x) \land Q(y) \to R(x,y))$
$\equiv \exists x\ \exists y\ \neg (\neg (P(x) \land Q(y)) \lor R(x,y))$
$\equiv \exists x\ \exists y\ (P(x) \land Q(y) \land \neg R(x,y))$

**EXAMPLE 2.3.7** Show that $\exists x\ (P(x) \land Q(x))$ is not equivalent to $\exists x\ P(x) \land \exists x\ Q(x)$.

*Proof*: We need to give an interpretation to verify that the two expresses have different truth values. Suppose the domain consists of all integers, $P(x)$ is $2x + 1 = 5$, and $Q(x)$ is $x^2 = 9$.

$\exists x\ (P(x) \land Q(x))$ is false, because there is no integer $a$ such that $2a + 1 = 5$ and $a^2 = 9$. However, there is an integer $b$ ($b = 2$), such that $2b + 1 = 5$, and there is an integer $c$ ($c = 3$ or $-3$) such that $c^2 = 9$. Therefore the statement $\exists x\ P(x) \land \exists x\ Q(x)$ is true. This counterexample is enough to show that $\exists x\ (P(x) \land Q(x))$ and $\exists x\ P(x) \land \exists x\ Q(x)$ are not equivalent.

## WORDS AND EXPRESSIONS

| | |
|---|---|
| interpretation | 解释 |
| valid | 有效的 |
| satisfiable | 可满足的 |
| substitution instance | 代换实例 |
| replacement rule | 代入规则 |
| renaming rule | 改名规则 |

## EXERCISES 2.3

1. Determine the truth values of the following wffs with an interpretation as follows:
   Let domain $D = \{1,2\}$, and the symbols in the wffs are assigned values as in the table:

   | $a$ | $b$ | $f(1)$ | $f(2)$ | $P(1,1)$ | $P(1,2)$ | $P(2,1)$ | $P(2,2)$ |
   |---|---|---|---|---|---|---|---|
   | 1 | 2 | 2 | 1 | 1 | 1 | 0 | 0 |

   a. $P(a,f(a)) \wedge P(b,f(b))$
   b. $\exists x\, P(x,f(a))$
   c. $\forall x\, \exists y\, P(y,x)$
   d. $\forall x\, \forall y\, (P(x,y) \rightarrow P(f(x),f(y)))$

2. Determine the truth values of the wffs.
   a. $\forall x\, (P(x) \rightarrow Q(f(x),a))$
   b. $\exists x\, (P(f(x)) \wedge Q(x,f(a)))$
   c. $\forall x\, \exists y\, (P(x) \wedge Q(x,y))$

   where the domain $D = \{1,2\}$, $a = 1$, and the values for the symbols are:

   | $f(1)$ | $f(2)$ | $P(1)$ | $P(2)$ | $Q(1,2)$ | $Q(2,1)$ | $Q(1,1)$ | $Q(2,2)$ |
   |---|---|---|---|---|---|---|---|
   | 2 | 1 | 0 | 1 | 1 | 0 | 1 | 0 |

3. Given an interpretation: the domain $D$ consists of all natural integers, $a = 2$, $f(x,y) = x+y$, $g(x,y) = xy$ and predicate $P(x,y)$ denotes the equality relation "$x = y$".
   Determine the truth values of the following wffs.
   a. $\forall x\, P(g(x,a),x)$
   b. $\forall x\, \forall y\, (P(f(x,a),y) \rightarrow P(f(y,a),x))$
   c. $\forall x\, \forall y\, \exists z\, P(f(x,y),z)$

d. $\exists x\, P(f(x,x), g(x,x))$

4. Rename the bound variables in the following wffs such that no bound variables use the same letter.
   a. $\forall x\, P(x,y,z) \to \exists y\, Q(x,y,z)$
   b. $\forall x\, \exists y\, (P(x,y) \to Q(y) \lor \forall y R(x,y,z))$
   c. $\forall x\, (P(x) \to (R(x) \lor Q(x))) \land \exists x\, R(x) \to \exists y\, S(x,y)$

5. Rename the free variables in the following wffs such that the free variable names are different from the bound variable names.
   a. $\forall x\, P(x,y) \to \exists y\, Q(x,y,z)$
   b. $(\exists y\, P(x,y) \to \forall x\, Q(x,z)) \land \exists x\, \forall z\, R(x,y,z)$
   c. $(\forall y\, P(x,y) \land \exists z\, Q(x,z)) \lor \forall x\, R(x,y)$

6. Show that $P(x,y) \to (Q(x,y) \to P(x,y))$ is valid.

7. Show that $\forall x\, (P(x) \to P(x)) \to \exists y\, (Q(y) \land \neg Q(y))$ is unsatisfiable.

8. Show that $\forall x\, \forall y\, (P(x,y) \to P(y,x))$ is satisfiable.

9. Show that the following equivalences hold.
   a. $\forall x\, P(x) \lor \forall x\, Q(x) \equiv \forall x\, \forall y\, (P(x) \lor Q(y))$
   b. $\forall x\, P(x) \land \exists x\, Q(x) \equiv \forall x\, \exists y\, (P(x) \land Q(y))$
   c. $\forall x\, \forall y\, (P(x) \to Q(y)) \equiv \exists x\, P(x) \to \forall y\, Q(y)$

## 2.4 Prenex Normal Form

In the propositional logic we know that any wff is equivalent to a wff in disjunctive normal form and to a wff in conjunctive normal form. Let's see whether we can do something similar with the wffs of the predicate logic. We will begin our discussion with a definition.

**DEFINITION 2.4.1** A wff $W$ is in **prenex normal form** if all of its quantifiers are on the left of the expression, namely, it is in the form of
$$Q_1 x_1 Q_2 x_2 \cdots Q_n x_n P(x_1, x_2, \cdots, x_n)$$
where each $Q_i\, (i = 1, 2, \cdots, n)$ is either $\forall$ or $\exists$, and $P(x_1, x_2, \cdots, x_n)$ is a predicate involving no quantifiers.

**EXAMPLE 2.4.1** $P(x)$, $\exists x\, P(x,y)$, $\forall x\, \exists y\, (P(x) \wedge Q(y))$, $\forall x\, \forall y\, \exists z\, (P(x) \vee Q(y) \wedge R(x,z))$, $\forall x\, P(x)$ are in prenex normal form, whereas $\exists x\, P(x) \vee \exists y\, Q(y)$ is not since the quantifiers do not all occur before the predicates.

In fact, every statement formed from propositional variables, predicates, $T$ and $F$ using logical connectives and quantifiers is equivalent to a wff in prenex normal form. There is an easy algorithm that can help us to obtain the desired form. The idea is to make sure that variables have distinct names and then apply equivalences that send all quantifiers to the left end of the wff. Here is the algorithm.

Any wff $W$ has an equivalent prenex normal form, which can be constructed as follows:
1. Rename the variables of $W$ such that no quantifiers use the same variable name and so that the bound variable names are distinct from the free variable names.
2. Move quantifiers to the left by using equivalences 4, 5 and 6 in Section 2.3.

The renaming of variables is very important to the success of the algorithm.

**EXAMPLE 2.4.2** Put the wffs in prenex normal form
1. $\forall x\, P(x) \wedge \neg \exists x\, Q(x)$
2. $\forall x\, P(x) \vee \neg \exists x\, Q(x)$
3. $\exists x\, P(x) \rightarrow \forall x\, Q(x)$
4. $\forall x\, P(x) \rightarrow \exists x\, Q(x)$

*Solution:*
1. $\quad \forall x\, P(x) \wedge \neg \exists x\, Q(x)$
$\equiv \forall x\, P(x) \wedge \neg \exists y\, Q(y)$
$\equiv \forall x\, P(x) \wedge \forall y\, \neg Q(y)$
$\equiv \forall x\, (P(x) \wedge \forall y\, \neg Q(y))$
$\equiv \forall x\, \forall y\, (P(x) \wedge \neg Q(y))$
2. $\quad \forall x\, P(x) \vee \neg \exists x\, Q(x)$
$\equiv \forall x\, P(x) \vee \neg \exists y\, Q(y)$
$\equiv \forall x\, P(x) \vee \forall y\, \neg Q(y)$
$\equiv \forall x\, \forall y\, (P(x) \vee \neg Q(y))$
3. $\quad \exists x\, P(x) \rightarrow \forall x\, Q(x)$
$\equiv \exists y\, P(y) \rightarrow \forall x\, Q(x)$
$\equiv \forall y\, (P(y) \rightarrow \forall x\, Q(x))$

$\equiv \forall y \ \forall x \ (P(y) \rightarrow Q(x))$

4. $\quad \forall x \ P(x) \rightarrow \exists x \ Q(x)$
$\equiv \forall y \ P(y) \rightarrow \exists x \ Q(x)$
$\equiv \exists y \ (P(y) \rightarrow \exists x \ Q(x))$
$\equiv \exists y \ \exists x \ (P(y) \rightarrow Q(x))$

**EXAMPLE 2.4.3** Put the wffs in prenex normal form

1. $(\forall x \ P(x,y) \rightarrow \exists y \ Q(y)) \rightarrow \forall x \ R(x,y,z)$
2. $P(x) \land \forall x \ (Q(x) \rightarrow \exists y \ R(x,y) \lor \neg \exists y \ P(y))$

*Solution:*

1. $\quad (\forall x \ P(x,y) \rightarrow \exists y \ Q(y)) \rightarrow \forall x \ R(x,y,z)$
$\equiv (\forall u \ P(u,y) \rightarrow \exists v \ Q(v)) \rightarrow \forall x \ R(x,y,z)$
$\equiv \exists u \ (P(u,y) \rightarrow \exists v \ Q(v)) \rightarrow \forall x \ R(x,y,z)$
$\equiv \exists u \ \exists v \ (P(u,y) \rightarrow Q(v)) \rightarrow \forall x \ R(x,y,z)$
$\equiv \forall u \ \forall v \ \forall x \ ((P(u,y) \rightarrow Q(v)) \rightarrow R(x,y,z))$

2. $\quad P(x) \land \forall x \ (Q(x) \rightarrow \exists y \ R(x,y) \lor \neg \exists y \ P(y))$
$\equiv P(x) \land \forall z \ (Q(z) \rightarrow \exists y \ R(z,y) \lor \forall y \ \neg P(y))$
$\equiv P(x) \land \forall z \ (Q(z) \rightarrow \exists u \ R(z,u) \lor \forall y \ \neg P(y))$
$\equiv P(x) \land \forall z \ \exists u \ \forall y \ (Q(z) \rightarrow R(z,u) \lor \neg P(y))$
$\equiv \forall z \ \exists u \ \forall y \ (P(x) \land (Q(z) \rightarrow R(z,u) \lor \neg P(y)))$

There are two special prenex normal forms that correspond to the disjunctive normal form and the conjunctive normal form for propositional calculus. We define a **literal** in the predicate calculus to be an atom or the negation of an atom. For example, $P(x)$ and $\neg Q(x,y)$ are literals. A prenex normal form is called a **prenex disjunctive normal form** if it has the form

$$Q_1 x_1 Q_2 x_2 \cdots Q_n x_n (D_1 \lor D_2 \lor \cdots \lor D_n)$$

where each $D_i$ is a conjunction of one or more literals. Similarly, a prenex normal form is called a **prenex conjunctive normal form** if it has the form

$$Q_1 x_1 Q_2 x_2 \cdots Q_n x_n (C_1 \land C_2 \land \cdots \land C_n)$$

where each $C_i$ is a disjunction of one or more literals.

It's easy to construct either of these normal forms from a prenex normal form by just eliminating conditionals, moving $\neg$ inwards, and either distributing $\land$ over $\lor$ or distributing $\lor$ over $\land$.

If we want to start with an arbitrary wff, we have the following nice algorithm.

Any wff $W$ has an equivalent prenex disjunctive/conjunctive normal form, which can be constructed as follows:

1. Rename the variables of $W$ such that no quantifiers use the same variable names and the quantified variable names are distinct from the free variable names;
2. Remove implications by using the equivalence $A \rightarrow B \equiv \neg A \lor B$;
3. Move negation to the right to form literals by using the equivalences $\neg \forall x \, A \equiv \exists x \, \neg A$ and $\neg \exists x \, A \equiv \forall x \, \neg A$, and De Morgan's laws;
4. Move quantifiers to the left to obtain a prenex normal form;
5. To obtain the disjunctive normal form, distribute $\land$ over $\lor$. To obtain the conjunctive normal form, distribute $\lor$ over $\land$.

**EXAMPLE 2.4.4** Put the wff $\forall x \, P(x) \lor \exists x \, Q(x) \rightarrow R(x) \land \exists x \, S(x)$ in prenex disjunctive normal form.

Solution:
$$\forall x \, P(x) \lor \exists x \, Q(x) \rightarrow R(x) \land \exists x \, S(x)$$
$$\equiv \forall y \, P(y) \lor \exists z \, Q(z) \rightarrow R(x) \land \exists u \, S(u)$$
$$\equiv \neg(\forall y \, P(y) \lor \exists z \, Q(z)) \lor (R(x) \land \exists u \, S(u))$$
$$\equiv (\exists y \, \neg P(y) \land \forall z \, \neg Q(z)) \lor (R(x) \land \exists u \, S(u))$$
$$\equiv \exists y \, \forall z \, (\neg P(y) \land \neg Q(z)) \lor \exists u \, (R(x) \land S(u))$$
$$\equiv \exists y \, \forall z \, \exists u \, ((\neg P(y) \land \neg Q(z)) \lor (R(x) \land S(u)))$$

**EXAMPLE 2.4.5** Put the wff $\forall x \, (P(x) \lor \forall z \, Q(z,y) \rightarrow \neg \forall y \, R(x,y))$ in prenex conjunctive normal form.

Solution:
$$\forall x (P(x) \lor \forall z \, Q(z,y) \rightarrow \neg \forall y \, R(x,y))$$
$$\equiv \forall x \, (P(x) \lor \forall z \, Q(z,y) \rightarrow \neg \forall w \, R(x,w))$$
$$\equiv \forall x \, (\neg(P(x) \lor \forall z \, Q(z,y)) \lor \neg \forall w \, R(x,w))$$
$$\equiv \forall x \, ((\neg P(x) \land \exists z \, \neg Q(z,y)) \lor \exists w \, \neg R(x,w))$$
$$\equiv \forall x \, (\exists z \, (\neg P(x) \land \neg Q(z,y)) \lor \exists w \, \neg R(x,w))$$
$$\equiv \forall x \, \exists z \, \exists w \, ((\neg P(x) \land \neg Q(z,y)) \lor \neg R(x,w))$$
$$\equiv \forall x \, \exists z \, \exists w \, ((\neg P(x) \lor \neg R(x,w)) \land (\neg Q(z,y) \lor \neg R(x,w)))$$

## WORDS AND EXPRESSIONS

prenex normal form           前束范式

prenex disjunctive normal form   前束析取范式
prenex conjunctive normal form   前束合取范式

## EXERCISES 2.4

1. Put the following statements in prenex normal form.
   a. $\exists x\, P(x) \lor \exists x\, Q(x) \lor A$, where $A$ is a proposition not involving any quantifiers
   b. $\neg(\forall x\, P(x) \lor \forall x\, Q(x))$
   c. $\exists x\, P(x) \rightarrow \exists x\, Q(x)$
   d. $\forall x\, (P(x) \rightarrow \exists y\, Q(x,y))$
   e. $\forall x\, \forall y\, (\exists z\, P(x,y,z) \land \exists u\, Q(x,u) \rightarrow \exists v\, Q(y,v))$

2. Construct a prenex disjunctive normal form for each of the following wffs.
   a. $\forall x\, (P(x) \lor Q(x)) \rightarrow \forall x\, P(x) \lor \forall x\, Q(x)$
   b. $\exists x\, P(x) \land \exists x\, Q(x) \rightarrow \exists x\, (P(x) \land Q(x))$
   c. $\forall x\, \exists y\, P(x,y) \rightarrow \exists y\, \forall x\, P(x,y)$
   d. $\forall x\, (P(x,f(x)) \rightarrow P(x,y))$
   e. $\forall x\, (P(x) \rightarrow \forall y\, (Q(x,y) \rightarrow R(y,x)))$
   f. $\forall x\, (P(x) \rightarrow Q(x,y)) \rightarrow (\exists y\, R(y) \rightarrow \exists z\, S(y,z))$

3. Construct a prenex conjunctive normal form for each of those wffs in Exercise 2.

## 2.5 Inference Rules in Predicate Calculus

To reason formally about wffs in the predicate calculus, we need some inference rules. It is nice to know that all inference rules of propositional calculus can still be used for the predicate logic. For example, let us take a look of the modus ponens rule of the propositional logic. If $A$ and $A \rightarrow B$ are valid wffs, then $B$ is valid. But sometimes, it is hard to reason about statements that contain quantifiers. The natural approach is to remove quantifiers from statements, do some reasoning with the unquantified statements, and then restore any needed quantifiers. But quantifiers can not be removed and restored at will. There are four rules of inference that govern the use of them.

### Universal Instantiation (UI)

It seems reasonable to say that if a property holds for everything, then it holds for any particular thing, that is, we should be able to infer $W(x)$ from $\forall x\ W(x)$. Similarly, we should be able to infer $W(c)$ from $\forall x\ W(x)$ whenever $c$ is a member of the domain. Therefore, we have:

$$\frac{\forall x\ W(x)}{\therefore W(x)} \quad \text{and} \quad \frac{\forall x\ W(x)}{\therefore W(c)}$$

where $c$ is a member of the domain of the wff $\forall x\ W(x)$.

**EXAMPLE 2.5.1** From the wff $\forall x\ (\forall y\ P(x,y) \lor Q(x))$ we can use UI to infer the following wffs.

$\forall y\ P(x,y) \lor Q(x)$
$\forall y\ P(c,y) \lor Q(c)$
$\forall y\ P(f(x,z,c),y) \lor Q(f(x,z,c))$

but we can not obtain $\forall y\ P(f(x,y),y) \lor Q(f(x,y))$.

## Existential Instantiation (EI)

It seems reasonable that whenever a property holds for some thing, then the property holds for a particular thing. Therefore, we have

$$\frac{\exists x\ W(x)}{\therefore W(c)}$$

where $c$ is a member of the domain such that $W(c)$ is true.

During proof, we must be careful about our choice of the constant $c$. For example, suppose $W(x) = P(x,c)$, can we infer $W(c)$ from $\exists x\ W(x)$? Let's look at an interpretation. Let $P(x,c) = $ "$x > c$" over all real numbers, where $c$ is a real number. Then $\exists x\ P(x,c)$ is true, but $P(c,c) = $ "$c > c$" is false. However, for our interpretation we know that $P(b,c) = $ "$b > c$" is true for some constant $b \neq c$. So we can infer $W(b)$ in this case. Therefore, we can infer $W(c)$ from $\exists x\ W(x)$ whenever $c$ is distinct from any constant in the wff $\exists x\ W(x)$.

Now, let's discuss another example. Suppose we have a statement:

$$\exists x\ P(x) \land \exists x\ Q(x).$$

Consider the following "attempted" proof:

1. $\exists x\ P(x) \land \exists x\ Q(x)$      Premise
2. $\exists x\ P(x)$      Simplification 1
3. $P(c)$      EI 2

## 2 Predicate Logic

> 4. $\exists x\, Q(x)$        Simplification 1
> 5. $Q(c)$        EI 4
> 6. $P(c) \wedge Q(c)$        Conjunction 3 and 5

If this proof is correct, then we have $\exists x\, P(x) \wedge \exists x\, Q(x) \rightarrow P(c) \wedge Q(c)$. But this wff is not valid. Consider the following interpretation: suppose $P(x) = $ "$x$ is female" and $Q(x) = $ "$x$ is male" over the domain of all people. Then $\exists x\, P(x) \wedge \exists x\, Q(x)$ is true, but $P(c) \wedge Q(c)$ is false for any person $c$. What went wrong? The problem was in line 5, where we used the same constant $c$ that had already been used in the proof. So each time using EI, we need to make sure a new constant is introduced. Therefore, we restate the existential instantiation rule as follows:

$$\frac{\exists x\, W(x)}{\therefore W(c)}$$

where $c$ is a new constant in the proof and an element of the domain, and $W(c)$ is true.

**EXAMPLE 2.5.2** Show that $\forall x\, \neg W(x) \rightarrow \neg \exists x\, W(x)$.

*Proof*:

> 1. $\neg(\neg \exists x\, W(x))$        Premise (negation of conclusion)
> 2. $\exists x\, W(x)$        Equivalent to 1
> 3. $W(c)$        EI 2
> 4. $\forall x\, \neg W(x)$        Premise
> 5. $\neg W(c)$        UI 4
> 6. $F$        Conjunction 3 and 5

## Universal Generalization (UG)

Universal generalization is the rule of inference stating that $\forall x\, W(x)$ is true, if $W(c)$ is true for any elements $c$ in the universe of discourse. Universal generalization is used when we show that $\forall x\, W(x)$ is true by taking an arbitrary element $c$ from the domain and showing that $W(c)$ is true. The element $c$ that we select must be arbitrary rather than a specific element of the domain:

$$\frac{W(c)}{\therefore \forall x\, W(x)}$$

where $c$ is an arbitrary element of the domain and $W(c)$ is true.

Sometimes, we have the following form:

$$\frac{W(x)}{\therefore \forall x\, W(x)}$$

where $W(x)$ is true for all elements of the domain.

**EXAMPLE 2.5.3** Show that $\forall x\, (P(x) \to Q(x)) \land \forall x\, (Q(x) \to R(x)) \to \forall x\, (P(x) \to R(x))$ is valid.

*Proof*:

1. $\forall x\, (P(x) \to Q(x))$      Premise
2. $P(x) \to Q(x)$      UI 1
3. $\forall x\, (Q(x) \to R(x))$      Premise
4. $Q(x) \to R(x)$      UI 3
5. $P(x) \to R(x)$      Hypothetical syllogism 2 and 4
6. $\forall x\, (P(x) \to R(x))$      UG 5

This argument is valid.

**EXAMPLE 2.5.4** (*Lewis Carroll's logic*) Show that the following reasoning is valid. Babies are illogical. Nobody is despised who can manage a crocodile. Illogical persons are despised. Therefore babies cannot manage crocodiles.

*Proof*: Let $P(x)$ mean "$x$ is a baby", $Q(x)$ "$x$ is logical", $R(x)$ "$x$ is despised", and $S(x)$ "$x$ can manage a crocodile". So the four sentences in the argument can be represented as follows:

The premises are: $\forall x\, (P(x) \to \neg Q(x))$, $\forall x\, (S(x) \to \neg R(x))$, $\forall x\, (\neg Q(x) \to R(x))$ and the conclusion is: $\forall x\, (P(x) \to \neg S(x))$.

1. $\forall x\, (P(x) \to \neg Q(x))$      Premise
2. $P(c) \to \neg Q(c)$      UI 1
3. $\forall x\, (S(x) \to \neg R(x))$      Premise
4. $S(c) \to \neg R(c)$      UI 3
5. $\forall x\, (\neg Q(x) \to R(x))$      Premise
6. $\neg Q(c) \to R(c)$      UI 5
7. $P(c) \to R(c)$      Hypothetical syllogism 2 and 6
8. $R(c) \to \neg S(c)$      Equivalent of 4
9. $P(c) \to \neg S(c)$      Hypothetical syllogism 7 and 8
10. $\forall x\, (P(x) \to \neg S(x))$      UG 9

## Existential Generalization (EG)

Existential generalization is the rule of inference that is used to conclude that $\exists x\, W(x)$ is true when a particular element $c$ with $W(c)$ is known true. That is, if we know one element $c$ in the universe of discourse for which $W(c)$ is true, then we know that $\exists x\, W(x)$ is true.

We state the existential generalization rule as follows:

$$\frac{W(c)}{\therefore \exists x\, W(x)}$$

$c$ is a particular element of the domain and $W(c)$ is true and $x$ does not occur in $W(c)$.

**EXAMPLE 2.5.5** Show that $\forall x\, P(x) \land \exists x\, Q(x) \rightarrow \exists x\, (P(x) \land Q(x))$ is valid.

*Proof*:

| | | |
|---|---|---|
| 1. | $\exists x\, Q(x)$ | Premise |
| 2. | $Q(c)$ | EI 1 |
| 3. | $\forall x\, P(x)$ | Premise |
| 4. | $P(c)$ | UI 3 |
| 5. | $P(c) \land Q(c)$ | Conjunction of 2 and 4 |
| 6. | $\exists x\, (P(x) \land Q(x))$ | EG 5 |

**EXAMPLE 2.5.6** Show that the premises "Everyone who majors in computer science is a logical thinker", and "John majors in computer science" imply the conclusion "There is some logical thinker".

*Proof*: Let $C(x)$ represent "$x$ majors in computer science", and $L(x)$ "$x$ is a logical thinker". Let constant $b$ be "John". Then the premises are $\forall x\, (C(x) \rightarrow L(x))$ and $C(b)$. The conclusion is $\exists x\, L(x)$.

| | | |
|---|---|---|
| 1. | $C(b)$ | Premise |
| 2. | $\forall x\, (C(x) \rightarrow L(x))$ | Premise |
| 3. | $C(b) \rightarrow L(b)$ | UI 2 |
| 4. | $L(b)$ | Modus Ponens 1 and 3 |
| 5. | $\exists x\, L(x)$ | EG 4 |

**EXAMPLE 2.5.7** Show that the premises "A student in this class has not done the homework", and "Everyone in this class passed the first exam" imply the conclusion "Someone who passed the first exam has not done the homework".

*Proof*: Let $C(x)$ be "$x$ is in this class", $H(x)$ be "$x$ has done the homework", and $P(x)$ be "$x$ passed the first exam". The premises are $\exists x\, (C(x) \wedge \neg H(x))$ and $\forall x\, (C(x) \rightarrow P(x))$. The conclusion is $\exists x\, (P(x) \wedge \neg H(x))$.

1. $\exists x\, (C(x) \wedge \neg H(x))$     Premise
2. $C(a) \wedge \neg H(a)$     EI 1
3. $\forall x\, (C(x) \rightarrow P(x))$     Premise
4. $C(a) \rightarrow P(a)$     UI 3
5. $C(a)$     Simplification 2
6. $P(a)$     Modus Ponens 4 and 5
7. $\neg H(a)$     Simplification 2
8. $P(a) \wedge \neg H(a)$     Conjunction of 6 and 7
9. $\exists x\, (P(x) \wedge \neg H(x))$     EG 8

**EXAMPLE 2.5.8** Show that $\exists x\, (P(x) \wedge Q(x)) \rightarrow \exists x\, P(x) \wedge \exists x\, Q(x)$ is valid.

*Proof*:

1. $\exists x\, (P(x) \wedge Q(x))$     Premise
2. $P(a) \wedge Q(a)$     EI 1
3. $P(a)$     Simplification 2
4. $\exists x\, P(x)$     EG 3
5. $Q(a)$     Simplification 2
6. $\exists x\, Q(x)$     EG 5
7. $\exists x\, P(x) \wedge \exists x\, Q(x)$     Conjunction of 4 and 6

## WORDS AND EXPRESSIONS

| | |
|---|---|
| universal instantiation | 全称量词消去 |
| existential instantiation | 存在量词消去 |
| universal generalization | 全称量词产生 |
| existential generalization | 存在量词产生 |

## EXERCISES 2.5

1. What rules of inference are used in the following famous argument?

    "All men are mortal. Socrates is a man. Therefore, Socrates is mortal."

2. For each of the following arguments, explain which rules of inference are used for each step.

   a. "Doug, a student in this class, knows how to write programs in Java. Everyone who knows how to write programs in Java can get a high-paying job. Therefore, someone in this class can get a high-paying job."

   b. "Somebody in this class enjoys whale watching. Every person who enjoys whale watching cares about ocean pollution. Therefore, there is a person in this class who cares about ocean pollution."

   c. "Each student in this class owns a personal computer. Everyone who owns a personal computer can use a word processing program. Therefore, John, a student in this class, can use a word processing program."

3. Show that each of the following wffs is valid.

   a. $\forall x\, (P(x) \to Q(x)) \land \exists x\, P(x) \to \exists x\, Q(x)$
   b. $\forall x\, (P(x) \to Q(x) \land R(x)) \land \forall x\, (P(x) \land S(x)) \to \forall x\, (R(x) \land S(x))$
   c. $\forall x\, (\neg P(x) \to Q(x)) \land \forall x\, \neg Q(x) \to \exists x\, P(x)$
   d. $\forall x\, (P(x) \lor Q(x)) \land \forall x\, (Q(x) \to \neg R(x)) \land \forall x\, R(x) \to \forall x\, P(x)$
   e. $\exists y\, \forall x\, P(x,y) \to \forall x\, \exists y\, P(x,y)$

4. Show that each of the following wffs is valid.

   a. $\forall x\, (P(x) \to Q(x)) \to (\exists x\, P(x) \to \exists x\, Q(x))$
   b. $\forall x\, (P(x) \to Q(x)) \to (\forall x\, P(x) \to \exists x\, Q(x))$
   c. $\forall x\, (P(x) \to Q(x)) \to (\forall x\, P(x) \to \forall x\, Q(x))$

# 3

# Set Theory

In the following we will study a wide variety of discrete structures. These include relations, which consist of ordered pairs of elements, and graphs, which are sets of vertices and edges connecting the vertices. Moreover, we will illustrate how these and other discrete structures are used in modeling and problem solving. In this chapter we study one of the most fundamental discrete structures called set, upon which all other discrete structures are built.

## 3.1 Sets

Formally a **set** is an unordered collection of objects called **elements** or **members** of the set. Sometimes the word **collection** is used in place of set to clarify a sentence. For example, "a collection of sets" is clearer than "a set of sets". We say that a set *contains* its elements, or that the elements *belong to* the set, or that the elements *are in* the set. If $S$ is a set and $x$ is an element in $S$, then we write $x \in S$. If $x$ is not an element of $S$, then we write $x \notin S$.

There are several ways to describe a set. One way is to list all the members of a set, if this is possible. We use the notation that all members of the set are listed between braces. For example, the notation $\{a, b, c, d, e\}$ represents the set with the five elements $a, b, c, d$ and $e$. Sometimes the brace notation is used to describe a set without listing all its members. In this situation some members of the set are listed, and then *ellipsis* ($\cdots$) is used when the general pattern of the elements is obvious. For example, the set $\{1, 2, 3, 4, 5, 6, 7, 8, 9, 10\}$ can be denoted in several different ways with ellipses, two of which are $\{1,2,3,\cdots,10\}$ and $\{1,2,3,\cdots,9,10\}$. The following sets, each denoted with a boldface letter, play important roles in discrete mathematics.

$\mathbf{Z} = \{0, 1, -1, 2, -2, 3, -3, \cdots\}$     the set of **integers**

$\mathbf{N} = \{0, 1, 2, 3, \cdots\}$     the set of **natural numbers**

$\mathbf{Z}^+ = \{1, 2, 3, \cdots\}$     the set of **positive integers**

$\mathbf{Q} = \{\frac{p}{q} : p \in \mathbf{Z}, q \in \mathbf{Z}, q \neq 0\}$     the set of **rational numbers**

$\mathbf{R}$     the set of **real numbers**

$\mathbf{C} = \{x + yi : x, y \in \mathbf{R}, i^2 = -1\}$     the set of **complex numbers**

$\mathbf{Z}_n = \{0, 1, 2, \cdots, n-1\}$, for any $n \in \mathbf{Z}^+$.

Also we have the following frequently used sets in mathematics: For $a, b \in \mathbf{R}$, $a < b$

$[a,b] = \{x \in \mathbf{R}, a \leqslant x \leqslant b\}$     the **closed interval**

$(a,b) = \{x \in \mathbf{R}, a < x < b\}$     the **open interval**

$(a,b] = \{x \in \mathbf{R}, a < x \leqslant b\}$     the (left) **half-open interval**

$[a,b) = \{x \in \mathbf{R}, a \leqslant x < b\}$     the (right) **half-open interval**

The set with no elements is called the **empty set**, or **null set**, and is denoted by $\{\}$ or more often by the symbol $\emptyset$. Often a set of elements with certain properties turns out to be the null set. For instance, the set of all positive integers that are greater than their squares is the null set. A set with one element is called a **singleton**. For example, $\{a\}, \{b\}, \{\emptyset\}$ are singletons.

Since many mathematical statements assert that two differently specified collections of objects are really the same set, we need to understand what it means for two sets to be equal.

Two sets $A$ and $B$ are equal if and only if they have the same elements. We denote the fact that $A$ and $B$ are equal sets by $A = B$. We can use equality to demonstrate two important characteristics of sets.

1. *There is no particular order or arrangement of the elements.*
2. *There are no redundant elements.*

For example, $\{a,b,c\} = \{c,a,b\} = \{a,b,c,c,c,a\}$.

Another way to describe a set is to use the **set builder** notation. We characterize all those elements in the set by stating the property or properties they must have to be members. For instance, the set $E$ of all even positive integers less than 10 can be written as:

$$E = \{x \mid x \text{ is an even positive integer less than } 10\}.$$

We often use this type of notation to describe sets when it is impossible to list all the elements of the set.

In dealing with sets, it's often useful to draw a picture to visualize the situation. A *Venn diagram*—named after the logician John Venn—consists of one or more closed curves in which the interior of each curve represents a set. Usually, the **universal set** $U$, which consists of all the objects under consideration, is represented by a rectangle. Inside this rectangle, circles or other geometrical figures are used to represent sets. Sometimes points are used to represent the particular elements of the set. Venn diagrams are often used to indicate the relationships between sets. For example, the Venn diagram in Figure 3.1.1 represents $V$, the set of vowels in the English alphabet.

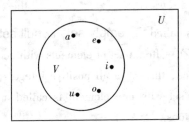

**Figure 3.1.1   Venn diagram for the set of vowels**

The rectangle indicates the universal set $U$ which is the set of the 26 letters of the English alphabet. Inside the rectangle, a circle represents $V$. Inside this circle, the points indicate the elements of $V$.

A set $A$ is said to be a subset of set $B$ if and only if every element of $A$ is also an element of $B$. We use the notation $A \subseteq B$ to indicate set $A$ is a subset of set $B$.

We see that $A \subseteq B$ if and only if the quantification $\forall x \ (x \in A \rightarrow x \in B)$ is true. It follows from the definition that every set $A$ is a subset of itself. Also, the empty set is a subset of any set $A$. So we have $\emptyset \subseteq A$. When we wish to emphasize that a set $A$ is a subset of set $B$ but $A \neq B$ we say that $A$ is a *proper subset* of $B$, and denote the fact by $A \subset B$. Thus,

# 3  Set Theory

$A \subset B$ if and only if $\forall x \ (x \in A \rightarrow x \in B) \land \exists x \ (x \in B \land x \notin A)$.

Sets are used extensively in counting problems, and for such applications we need to discuss the size of sets. Let $S$ be a set, if there are exactly $n$ distinct elements in $S$ where $n$ is a nonnegative integer, we say that $S$ is a **finite** set and that $n$ is the **cardinality** of $S$. The cardinality of $S$ is denoted by $|S|$. For example let $S$ be the set of letters in the English alphabet. Then $|S| = 26$. Since the null set has no elements, it follows that $|\varnothing| = 0$. Let $A = \{0, 1, 2, \{a, b\}, a\}$, then $|A| = 5$, not 6.

A set is said to be **infinite** if it is not finite. The collection of all subsets of a set $S$ is called the **power set** of $S$. The power set of $S$ is denoted by $P(S)$. For example if $S = \{1, 2, 3\}$, then the power set of $S$ can be written as $P(S) = \{\varnothing, \{1\}, \{2\}, \{3\}, \{1, 2\}, \{1, 3\}, \{2, 3\}, \{1, 2, 3\}\}$, and $|P(S)| = 8$.

In general, for any finite set $A$ with $|A| = n \geqslant 0$, we find that $A$ has $2^n$ subsets and therefore $|P(A)| = 2^n$. For any $0 \leqslant k \leqslant n$, there are $C_n^k$ subsets of size $k$. Counting the subsets of $A$ containing $k$ elements, we obtain the combinatorial identity $C_n^0 + C_n^1 + \cdots + C_n^n = 2^n$, for $n \geqslant 0$.

## WORDS AND EXPRESSIONS

| collection, set | 集合 |
| integer | 整数 |
| natural number | 自然数 |
| real number | 实数 |
| rational number | 有理数 |
| complex number | 复数 |
| closed interval | 闭区间 |
| open interval | 开区间 |
| singleton | 单元集 |
| Venn diagram | 文氏图 |
| universal set | 全集 |
| subset | 子集 |
| proper subset | 真子集 |
| cardinality | 基数 |
| power set | 幂集 |

# EXERCISES 3.1

1. List the members of the following sets.
   a. $\{x \mid x$ is a real number such that $x^2 = 1\}$
   b. $\{x \mid x$ is a positive integer less than $x^2 = 1\}$
   c. $\{x \mid x$ is the square of an integer and $x < 50\}$
   d. $\{x \mid x$ is an integer such that $x^2 = 2\}$

2. Determine whether the following pairs of sets are equal.
   a. $\{1, 3, 3, 5, 5\}$, $\{5, 3, 1\}$
   b. $\{\{1\}\}$, $\{1, \{1\}\}$
   c. $\emptyset$, $\{\emptyset\}$

3. Let $A = \{1, \{1\}, \{2\}\}$, which of the following statements are true?
   $1 \in A$;   $\{1\} \in A$;   $\{1\} \subseteq A$;   $\{\{1\}\} \subseteq A$;
   $2 \in A$;   $\{2\} \in A$;   $\{2\} \subseteq A$;   $\{\{2\}\} \subseteq A$.

4. For $A = \{1,2,3,4,5,6,7\}$, determine the number of
   a. subsets of $A$
   b. nonempty subsets of $A$
   c. proper subsets of $A$
   d. nonempty proper subsets of $A$
   e. subsets of $A$ containing three elements
   f. subsets of $A$ containing 1 and 2
   g. subsets of $A$ containing five elements, including 1 and 2
   h. proper subsets of $A$ containing 1 and 2

5. Answer questions:
   a. If a set $A$ has 63 proper subsets, what is $|A|$?
   b. If a set $B$ has 64 subsets of odd cardinality, what is $|B|$?
   c. Generalize the result of part b.

6. Let $S = \{1,2,3, \cdots, 29, 30\}$. How many subsets $A$ of $S$ satisfy
   a. $|A| = 5$?
   b. $|A| = 5$ and the smallest element in $A$ is 5?

7. Give an example of three sets $A, B$ and $C$ such that $A \in B$ and $B \in C$ but $A \notin C$.

8. Show that $\emptyset \subset A$ for every set $A$.

9. Write down the power set for each of the following sets.
   $\{a,b,c\}$;   $\{a,\{a,b\}\}$;   $\{\{a\}\}$;   $\{\{a\},\{\{b\}\}\}$

10. How many elements does each of these sets have?
    $P(\{a,b,\{a,b\}\})$;   $P(\{\emptyset, a, \{a\}, \{\{a\}\}\})$

## 3.2 Set Operations

Two sets can be combined in many different ways. We now introduce the following operations for sets.

**DEFINITION 3.2.1** Let $U$ be the universal set. For $A, B \subseteq U$, we define the following operation.

1. The **union** of $A$ and $B$, $A \cup B = \{x \mid x \in A \lor x \in B\}$
2. The **intersection** of $A$ and $B$, $A \cap B = \{x \mid x \in A \land x \in B\}$
3. The **difference** of $A$ and $B$, $A - B = \{x \mid x \in A \land x \notin B\}$
4. The **symmetric difference** of $A$ and $B$, $A \Delta B = \{x \mid x \in A \cup B \land x \notin A \cap B\}$
5. The **complement** of $A$, $\overline{A} = U - A = \{x \mid x \notin A\}$

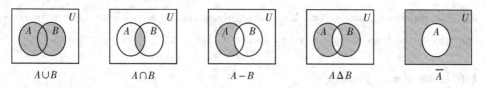

Figure 3.2.1  Set operations

Note that if $A, B \subseteq U$, then $A \cup B, A \cap B, A - B, A \Delta B \subseteq U$. Consequently, $\cup, \cap, -,$ and $\Delta$ are closed operations on $P(U)$, and we may also say that $P(U)$ is closed under these operations. The operations are demonstrated in Figure 3.2.1 where the resulted sets are shaded.

**EXAMPLE 3.2.1**  With $U = \{1, 2, 3, \cdots, 9, 10\}, A = \{1, 2, 3, 4, 5\}, B = \{3, 4, 5, 6, 7\}$ and $C = \{7, 8, 9\}$, we have

1. $A \cup B = \{1, 2, 3, 4, 5, 6, 7\}$
2. $A \cap B = \{3, 4, 5\}$
3. $A \cup C = \{1, 2, 3, 4, 5, 7, 8, 9\}$

4. $A \cap C = \emptyset$
5. $A - B = \{1,2\}$
6. $B - A = \{6,7\}$
7. $A \Delta C = A \cup C - A \cap C = \{1,2,3,4,5,7,8,9\} - \emptyset = \{1,2,3,4,5,7,8,9\}$
8. $A \Delta B = A \cup B - A \cap B = \{1,2,6,7\}$
9. $\overline{A} = U - A = \{6,7,8,9,10\}$

**DEFINITION 3.2.2** Two sets are **disjoint** if their intersection is empty set.

**EXAMPLE 3.2.2** Show that if $A, B \subseteq U$, then $A$ and $B$ are disjoint if and only if $A \cup B = A \Delta B$.

*Proof*: We start with $A$ and $B$ being disjoint. Consider any $x \in U$. If $x \in A \cup B$, then $x \in A$ or $x \in B$. But $A \cap B = \emptyset$, $x \notin A \cap B$ so $x \in A \Delta B$. Because $x \in A \cup B$ implies $x \in A \Delta B$, consequently we have $A \cup B \subseteq A \Delta B$. For the opposite inclusion, if $y \in A \Delta B$, then $y \in A$ or $y \in B$. So $y \in A \cup B$, Therefore $A \Delta B \subseteq A \cup B$. We have $A \Delta B = A \cup B$.

We prove the converse by contradiction. Consider any $A, B \subseteq U$ satisfying $A \cup B = A \Delta B$. If $A \cap B \neq \emptyset$, then there is $x \in A \cap B$, i. e., $x \in A$ and $x \in B$. Hence $x \in A \cup B$ ( = $A \Delta B$). But $x \in A \cup B$ and $x \in A \cap B$, so $x \notin A \Delta B$. From this contradiction we conclude that $A$ and $B$ are disjoint.

We now introduce some major laws that govern set theory. These bear a marked similarity to the laws of logic given in Section 1.3.

## The laws of set Theory

For any sets $A$, $B$, and $C$ taken from a universal set $U$,

1. $\overline{\overline{A}} = A$      Double complementation law
2. $A \cup \emptyset = A$      Identity laws
   $A \cap U = A$
3. $A \cup U = U$      Domination laws
   $A \cap \emptyset = \emptyset$
4. $A \cup A = A$      Idempotent laws
   $A \cap A = A$
5. $A \cup B = B \cup A$      Commutative laws
   $A \cap B = B \cap A$

6. $A \cup (B \cup C) = (A \cup B) \cup C$      Associative laws
$A \cap (B \cap C) = (A \cap B) \cap C$
7. $A \cap (B \cup C) = (A \cap B) \cup (A \cap C)$      Distributive laws
$A \cup (B \cap C) = (A \cup B) \cap (A \cup C)$
8. $\overline{A \cup B} = \overline{A} \cap \overline{B}$      De Morgan's laws
$\overline{A \cap B} = \overline{A} \cup \overline{B}$
9. $A \cup (A \cap B) = A$      Absorption laws
$A \cap (A \cup B) = A$
10. $A \cup \overline{A} = U$      Inverse laws
$A \cap \overline{A} = \emptyset$

**EXAMPLE 3.2.3** Prove that $\overline{A \cap B} = \overline{A} \cup \overline{B}$.

*Proof:* We will prove the equality by showing that each is a subset of the other.

First, suppose that $x \in \overline{A \cap B}$. Then $x \notin A \cap B$. Hence $\neg (x \in A \wedge x \in B)$ is true. Applying De Morgan's law, we have $x \notin A \vee x \notin B$, that is, $x \in \overline{A}$ or $x \in \overline{B}$. It follows that $x \in \overline{A} \cup \overline{B}$. This shows that $\overline{A \cap B} \subseteq \overline{A} \cup \overline{B}$.

Now suppose that $x \in \overline{A} \cup \overline{B}$, then $x \in \overline{A}$ or $x \in \overline{B}$. Hence $x \notin A$ or $x \notin B$. Consequently $\neg (x \in A) \vee \neg (x \in B)$ is true. It follows that $\neg (x \in A \cap B)$ holds. We use the definition of complement to conclude that $x \in \overline{A \cap B}$. This shows that $\overline{A} \cup \overline{B} \subseteq \overline{A \cap B}$.

Since we have shown that each set is a subset of the other, the two sets are equal, and the identity is proved.

**EXAMPLE 3.2.4** Use set builder notation and logical equivalences to show that $\overline{A \cap B} = \overline{A} \cup \overline{B}$.

*Solution:* $\overline{A \cap B} = \{x \mid x \notin A \cap B\} = \{x \mid \neg (x \in A \wedge x \in B)\} = \{x \mid (x \notin A \vee x \notin B)\}$
$= \{x \mid (x \in \overline{A} \vee x \in \overline{B})\} = \{x \mid x \in (\overline{A} \cup \overline{B})\} = \overline{A} \cup \overline{B}$

**EXAMPLE 3.2.5** Let $A$, $B$, and $C$ be sets. Show that $\overline{A \cup (B \cap C)} = (\overline{C} \cup \overline{B}) \cap \overline{A}$.

*Solution:* $\overline{A \cup (B \cap C)} = \overline{A} \cap \overline{(B \cap C)} = \overline{A} \cap (\overline{B} \cup \overline{C}) = (\overline{C} \cup \overline{B}) \cap \overline{A}$.

**EXAMPLE 3.2.6** Express $A - B$ in terms of $\cup$ and $\overline{\phantom{x}}$.

*Solution:* From the definition of difference, $A - B = \{x \mid x \in A \wedge x \notin B\} = A \cap \overline{B}$. Therefore, $A - B = A \cap \overline{B} = \overline{\overline{A} \cup \overline{\overline{B}}} = \overline{\overline{A} \cup B}$.

One more technique for establishing set equalities is the **membership table**. (This method is alike of using the truth table in Section 1.1.) We observe that for sets $A, B \subseteq U$, an element $x \in U$ satisfies exactly one of the following four situations:

1. $x \notin A, x \notin B$
2. $x \notin A, x \in B$
3. $x \in A, x \notin B$
4. $x \in A, x \in B$

When $x$ is an element of a given set, we write a 1 in the column representing that set in the membership table; When $x$ is not in the set, we enter a 0. Table 3.2.1 gives the membership table for $A \cap B$ and $A \cup B$. For example, the second row of the table tells us that when an element $x \in U$ is in the set $B$ but not in $A$, then it is not in $A \cap B$ but it is in $A \cup B$.

**Table 3.2.1** Membership table for $A \cap B$ and $A \cup B$

| A | B | $A \cap B$ | $A \cup B$ |
|---|---|---|---|
| 0 | 0 | 0 | 0 |
| 0 | 1 | 0 | 1 |
| 1 | 0 | 0 | 1 |
| 1 | 1 | 1 | 1 |

Using membership tables, we can establish the equality of two sets by comparing their respective columns in the table.

**EXAMPLE 3.2.7** Use a membership table to show that $A \cap (B \cup C) = (A \cap B) \cup (A \cap C)$.

Solution: We establish the membership table as Table 3.2.2.

**Table 3.2.2** Membership table for $A \cap (B \cup C)$ and $(A \cap B) \cup (A \cap C)$

| A | B | C | $B \cup C$ | $A \cap (B \cup C)$ | $A \cap B$ | $A \cap C$ | $(A \cap B) \cup (A \cap C)$ |
|---|---|---|---|---|---|---|---|
| 0 | 0 | 0 | 0 | 0 | 0 | 0 | 0 |
| 0 | 0 | 1 | 1 | 0 | 0 | 0 | 0 |
| 0 | 1 | 0 | 1 | 0 | 0 | 0 | 0 |
| 0 | 1 | 1 | 1 | 0 | 0 | 0 | 0 |
| 1 | 0 | 0 | 0 | 0 | 0 | 0 | 0 |
| 1 | 0 | 1 | 1 | 1 | 0 | 1 | 1 |
| 1 | 1 | 0 | 1 | 1 | 1 | 0 | 1 |
| 1 | 1 | 1 | 1 | 1 | 1 | 1 | 1 |

This table has eight rows. Since the columns for $A \cap (B \cup C)$ and $(A \cap B) \cup (A \cap C)$ are the same, the identity is valid.

In closing this section we introduce the computer representation of sets. There are various ways to represent sets using a computer. One method is to store the elements of a set in an unordered fashion. However, if this is done, the operations of computing the union, intersection, or difference of two sets would be time-consuming, since each of these operations would require a large amount of searching for elements. We will present a method for storing elements using an arbitrary ordering of the elements of the universal set. This method of representing sets makes the operations of sets easy.

Assume that the universal set $U$ is finite ( and of reasonable size so that the memory size required to store the elements of $U$ is not larger than the memory size of the computer being used). First, specify an arbitrary ordering of the elements of $U$, for instance $a_1$, $a_2$, $\cdots$, $a_n$. Represent a subset $A$ of $U$ with a bit string of length $n$, where the $i$ th bit in this string is 1 if $a_i$ belongs to $A$ and is 0 if $a_i$ does not belong to $A$. EXAMPLE 3.2.8 illustrates this technique.

**EXAMPLE 3.2.8** Let $U = \{1, 2, 3, 4, 5, 6, 7, 8\}$ and the ordering of the elements of $U$ have the elements in increasing order, i.e., $a_i = i$. What bit strings represent the subset $A = \{2, 3, 6, 7\}$, the subset of all even integers in $U$ and the subset of integers not exceeding 5 in $U$?

*Solution*: The bit string that represents $A = \{2,3,6,7\}$ has a 1 bit in the second, third, sixth and seventh positions, and a 0 elsewhere. It is 01100110. The set of all even integers in $U$, namely, $\{2,4,6,8\}$, is represented by the string 01010101. Similarly, we represent the set of integers not exceeding 5 in $U$, namely, $\{1,2,3,4,5\}$, by the string 11111000.

Representing sets by bit strings makes it easy to find the complements of sets, as well as the unions, intersections, and differences of sets. To find the bit string for the complement of a set from the bit string for that set, we simply change each 1 to 0 and each 0 to 1, since $x \in A$ if and only if $x \notin \bar{A}$. Note that this operation corresponds to taking the negation of each bit when we associate a bit with a truth value 1 representing true and 0 representing false.

**EXAMPLE 3.2.9** What is the bit string for the complement of the set $A = \{2,3,6,7\}$

with the universal set $U = \{1,2,3,4,5,6,7,8\}$?

*Solution*: EXAMPLE 3.2.8 shows that the bit string for $A$ is 01100110. The bit string for $\overline{A}$ is obtained by replacing 0s with 1s and vice versa. This yields the string 10011001, which corresponds to the set $\overline{A} = \{1,4,5,8\}$.

To obtain the bit string for the union and intersection of two sets we perform bitwise Boolean operations on the bit strings representing the two sets. The bit in the $i$th position of the bit string of the union is 1 if either of the bits in the $i$th position in the two strings is 1 (or both are 1), and is 0 if both bits are 0. The bit in the $i$th position of the bit string of the intersection is 1 if both bits in the corresponding position in the two strings are 1, and is 0 when either of the two bits is 0 (or both are 0).

**EXAMPLE 3.2.10** The bit strings of the sets $\{1,2,3,4,5\}$ and $\{2,3,6,7\}$ are 11111000 and 01100110, respectively. Use bit strings to find the union and intersection of these sets.

*Solution*: The bit string for the union of these sets is $11111000 \vee 01100110 = 11111110$, which corresponds to the set $\{1,2,3,4,5,6,7\}$. The bit string for the intersection of these sets is $11111000 \wedge 01100110 = 01100000$, which corresponds to the set $\{2,3\}$.

## WORDS AND EXPRESSIONS

| | |
|---|---|
| union | 并 |
| intersection | 交 |
| difference | 差 |
| symmetric difference | 对称差 |
| complement | 补 |
| membership table | 从属关系表 |

## EXERCISES 3.2

1. Let $U = \{1,2,3,\cdots,10\}$, $A = \{1,2,3,4,5\}$, $B = \{1,2,4,8\}$, $C = \{1,2,3,5,7\}$ and $D = \{2,4,6,8\}$. Find

   $(A \cup B) \cap C$      $\overline{C \cap D}$      $(B - C) - D$

   $A \cup (B \cap C)$      $(A \cup B) - C$      $B - (C - D)$

   $C \triangle D$      $A \cup (B - C)$      $(A \cup B) - (C \cap D)$

2. Given $A - B = \{1,3,7,11\}$, $B - A = \{2,6,8\}$ and $A \cap B = \{4,9\}$, determine the sets $A$ and $B$.

3. Let $C - D = \{1,2,4\}$, $D - C = \{7,8\}$ and $C \cup D = \{1,2,4,5,7,8,9\}$. Find $C$ and $D$.

4. Let $A$ and $B$ be sets. Show that:
$(A \cap B) \subseteq A$ $\qquad A \subseteq A \cup B \qquad A - B \subseteq A$
$A \cap (B - A) = \emptyset \qquad A \cup (B - A) = A \cup B \qquad (A \cap B) \cup (A \cap \overline{B}) = A$

5. $A$, $B$ and $C$ are sets, show that:
   a. $A \cup B \subseteq A \cup B \cup C$
   b. $A \cap B \cap C \subseteq A \cap B$
   c. $(A - B) - C \subseteq A - C$
   d. $(A - C) \cap (C - B) = \emptyset$
   e. $(B - A) \cup (C - A) = (B \cup C) - A$
   f. $(A - B) - C = (A - C) - (B - C)$

6. Prove or disprove each of the following for sets $A, B \subseteq U$
   a. $P(A \cup B) = P(A) \cup P(B)$
   b. $P(A \cap B) = P(A) \cap P(B)$

7. Use membership tables to establish each of the following equalities:
   a. $\overline{A \cap B} = \overline{A} \cup \overline{B}$
   b. $A \cup (A \cap B) = A$
   c. $(A \cap B) \cup (\overline{A} \cap C) = (A \cap \overline{B}) \cup (\overline{A} \cap \overline{C})$

8. Using the laws of set theory simplify each of the following:
   a. $((A \cup B) \cap B) - (A \cup B)$
   b. $((A \cup B \cup C) - (B \cup C)) \cup A$
   c. $(B - (A \cap C)) \cup (A \cap B \cap C)$
   d. $(A \cap B) - (C - (A \cup B))$
   e. $(A - B) \cup (A \cap C)$

9. Suppose that the universal set is $U = \{1,2,\cdots,9,10\}$. Express each of these sets with bit strings where the $i$th bit in the string is 1 if $i$ is in the set and 0 otherwise.
$\{1,3,4,5\} \quad \{2,3,6,8\} \quad \{3,4,5,6,7,8\}$

10. Using the same universal set as in the last problem, find the set specified by each of the bit strings.
1100100100   0101011011

## 3.3  Inclusion-Exclusion

Often, we are interested in finding the cardinality of the union of sets. To find the number of elements in the union of two finite sets $A$ and $B$, note that $|A| + |B|$ counts each element that is in $A$ but not in $B$ or in $B$ but not in $A$ exactly once, and each element that is in both $A$ and $B$ exactly twice. Thus, if the number of elements that are in both $A$ and $B$ is subtracted from $|A| + |B|$, elements in $A \cap B$ will be counted only once. Hence,

$$|A \cup B| = |A| + |B| - |A \cap B|.$$

**EXAMPLE 3.3.1**  A discrete mathematics class contains 97 students majoring in computer science, 20 students majoring in mathematics, and 10 joint mathematics and computer science majors. How many students are in this class, if every student is majoring in mathematics, computer science or both mathematics and computer science?
*Solution*: Let $A$ be the set of students in the class majoring in computer science and $B$ be the set of students in the class majoring in mathematics. $A \cap B$ is the set of students in the class who are joint mathematics and computer science majors. Since every student in the class is majoring in either computer science or mathematics, it follows that the number of students in the class is $|A \cup B| = |A| + |B| - |A \cap B| = 97 + 20 - 10$. Therefore, there are 107 students in the class.

**EXAMPLE 3.3.2**  How many positive integers not exceeding 1000 are divisible by 9 or 11?
*Solution*: Let $A$ be the set of positive integers not exceeding 1000 that are divisible by 9, and $B$ be the set of positive integers not exceeding 1000 that are divisible by 11. Then $A \cup B$ is the set of positive integers not exceeding 1000 that are divisible by either 9 or 11, and $A \cap B$ is the set of positive integers not exceeding 1000 that are divisible by both 9 and 11. There are $\left\lfloor \frac{1000}{9} \right\rfloor$ positive integers divisible by 9 and $\left\lfloor \frac{1000}{11} \right\rfloor$ positive integers divisible by 11, and $\left\lfloor \frac{1000}{9 \times 11} \right\rfloor$ positive integers divisible by both 9 and 11. It

follows that there are $|A \cup B| = |A| + |B| - |A \cap B| = \left|\dfrac{1000}{9}\right| + \left|\dfrac{1000}{11}\right| - \left|\dfrac{1000}{9 \times 11}\right| = 191$ positive integers not exceeding 1000 that are divisible by either 9 or 11.

This calculation can be extended to three sets. To count the number of the elements in the union of three sets $A$, $B$ and $C$, note that $|A| + |B| + |C|$ counts each element that is in exactly one of the three sets once, the elements that are in exactly two of the sets twice, and the elements in all three sets three times. To remove the counting of the elements in more than one of the sets, subtract the number of elements in the intersections of all pairs of the three sets, obtaining

$$|A| + |B| + |C| - |A \cap B| - |A \cap C| - |B \cap C|.$$

This expression still counts the elements that occur in exactly one of the sets once. An element that occurs exactly in two of the sets is also counted exactly once, since this element will occur in one of the three intersections of the sets. However, those elements that occur in all three sets will be counted zero times by this expression, since they occur in all three intersections of two sets. To remedy this undercount, add the number of elements in the intersection of all three sets. The final expression counts each element once, whether it is in one, two, or three of the sets. Thus,

$$|A \cup B \cup C| = |A| + |B| + |C| - |A \cap B| - |A \cap C| - |B \cap C| + |A \cap B \cap C|.$$

**EXAMPLE 3.3.3** How many positive integers not exceeding 1000 are not divisible by 5, 6, and 8?

*Solution*: Let $A$, $B$ and $C$ be the sets of positive integers not exceeding 1000 that are divisible by 5, 6 and 8, respectively. Then $|A \cup B \cup C|$ is the set of positive integers not exceeding 1000 that are divisible by 5, 6, or 8. $\overline{A} \cap \overline{B} \cap \overline{C}$ is the set of positive integers not exceeding 1000 that are not divisible by 5, 6 and 8. From De Morgan's laws, $\overline{A} \cap \overline{B} \cap \overline{C} = \overline{A \cup B \cup C}$, we have that $|\overline{A} \cap \overline{B} \cap \overline{C}| = |\overline{A \cup B \cup C}| = |U| - |A \cup B \cup C|$. We know that $|U| = 1000$. Therefore,

$$|\overline{A} \cap \overline{B} \cap \overline{C}| = 1000 - |A| - |B| - |C| + |A \cap B| + |A \cap C| + |B \cap C| - |A \cap B \cap C|$$

Let $lcm(x_1, x_2, \cdots, x_n)$ denotes the least common multiple of positive integers $x_1, x_2, \cdots, x_n$. Then

$$|A| = \left|\dfrac{1000}{5}\right| = 200, \qquad |B| = \left|\dfrac{1000}{6}\right| = 166,$$

$$|C| = \left|\frac{1000}{8}\right| = 125, \qquad |A \cap B| = \left|\frac{1000}{5 \times 6}\right| = 33,$$

$$|B \cap C| = \left|\frac{1000}{lcm(6,8)}\right| = 41, \qquad |A \cap C| = \left|\frac{1000}{5 \times 8}\right| = 25,$$

$$|A \cap B \cap C| = \left|\frac{1000}{lcm(5,6,8)}\right| = 8.$$

Thus,
$$|\overline{A} \cap \overline{B} \cap \overline{C}| = 1000 - 200 - 166 - 125 + 33 + 41 + 25 - 8 = 600.$$

Therefore, there are 600 positive integers not exceeding 1000 that are not divisible by 5, 6, and 8.

**EXAMPLE 3.3.4** A total of 1300 students have taken Advanced Mathematics, 800 have taken Advanced Algebra, and 200 have taken Advanced Geometry, 100 have taken both Advanced Mathematics and Advanced Algebra, 50 have taken both Advanced Mathematics and Advanced Geometry, and 30 have taken both Advanced Algebra and Advanced Geometry. If 2500 students have taken at least one of Advanced Mathematics, Advanced Algebra and Advanced Geometry, how many students have taken all three courses?

*Solution*: Let $M$ be the set of students who have taken Advanced Mathematics, $A$ the set of students who have taken Advanced Algebra, and $G$ the set of students who have taken Advanced Geometry. Then

$$|M| = 1300, \ |A| = 800, \ |G| = 200,$$
$$|M \cap A| = 100, \ |M \cap G| = 50, \ |A \cap G| = 30 \text{ and } |M \cup A \cup G| = 2500.$$

Substituting these quantities into the equation

$$|M \cup A \cup G| = |M| + |A| + |G| - |M \cap A| - |M \cap G| - |A \cap G| + |M \cap A \cap G|$$

results in

$$2500 = 1300 + 800 + 200 - 100 - 50 - 30 + |M \cap A \cap G|.$$

Hence $|M \cap A \cap G| = 380$. Therefore, there are 380 students who have taken Advanced Mathematics, Advanced Algebra and Advanced Geometry.

We will now state and prove the inclusion-exclusion principle, which tells us how many elements are in the union of a finite number of finite sets.

**THEOREM 3.3.1** (The Principle of Inclusion-Exclusion) Let $A_1, A_2, \cdots, A_n$ be finite sets. Then

$$|A_1 \cup A_2 \cup \cdots \cup A_n| = \sum |A_i| - \sum |A_i \cap A_j|$$
$$+ \sum_{1 \leq i,j,k \leq n} |A_i \cap A_j \cap A_k| - \cdots + (-1)^{n+1}|A_1 \cap A_2 \cap \cdots \cap A_n|.$$

*Proof*: We will prove the formula by showing that an element in the union is counted exactly once by the right-hand side of the equation. Suppose that $a$ is exactly a member of $r$ sets of the sets $A_1, A_2, \cdots, A_n$ for $1 \leq r \leq n$. This element is counted $C_r^1$ times by $\sum |A_i|$, and $C_r^2$ times by $\sum |A_i \cap A_j|$. In general, it is counted $C_r^m$ times by the summation involving $m$ of the sets of $A_i$. Thus, this element is counted exactly

$$C_r^1 - C_r^2 + C_r^3 - \cdots + (-1)^{r+1} C_r^r$$

times by the expression on the right-hand side of this equation. To evaluate this quantity, we have

$$C_r^0 - C_r^1 + C_r^2 - \cdots + (-1)^r C_r^r = 0.$$

Hence,

$$1 = C_r^0 = C_r^1 - C_r^2 + C_r^3 - \cdots + (-1)^{r+1} C_r^r.$$

Therefore, each element in the union is counted exactly once by the expression on the right-hand side of the equation. This proves the principle of inclusion-exclusion.

**EXAMPLE 3.3.5** How many integers between 1 and 250 (inclusive) that are divisible by at least one of 2, 3, 5 and 7?

*Solution*: Let $A, B, C$, and $D$ be sets of positive integers not exceeding 250 that are divisible by 2, 3, 5, and 7, respectively. Then

$$|A| = \left|\frac{250}{2}\right| = 125, \qquad |B| = \left|\frac{250}{3}\right| = 83,$$

$$|C| = \left|\frac{250}{5}\right| = 50, \qquad |D| = \left|\frac{250}{7}\right| = 35,$$

$$|A \cap B| = \left|\frac{250}{2 \times 3}\right| = 41, \qquad |A \cap C| = \left|\frac{250}{2 \times 5}\right| = 25,$$

$$|A \cap D| = \left|\frac{250}{2 \times 7}\right| = 17, \qquad |B \cap C| = \left|\frac{250}{3 \times 5}\right| = 16,$$

$$|B \cap D| = \left|\frac{250}{3 \times 7}\right| = 11, \qquad |C \cap D| = \left|\frac{250}{5 \times 7}\right| = 7,$$

$$|A \cap B \cap C| = \left|\frac{250}{2 \times 3 \times 5}\right| = 8, \qquad |A \cap B \cap D| = \left|\frac{250}{2 \times 3 \times 7}\right| = 5,$$

$$|A \cap C \cap D| = \left|\frac{250}{2 \times 5 \times 7}\right| = 3, \qquad |B \cap C \cap D| = \left|\frac{250}{3 \times 5 \times 7}\right| = 2,$$

$$|A \cap B \cap C \cap D| = \left| \frac{250}{2 \times 3 \times 5 \times 7} \right| = 1.$$

Hence we have

$|A \cup B \cup C \cup D| = 125 + 83 + 50 + 35 - 41 - 25 - 17 - 16 - 11 - 7 + 8 + 5 + 3 + 2 - 1 = 193.$

Therefore, there are 193 integers not exceeding 250 that are divisible by at least one of 2, 3, 5, and 7.

## WORDS AND EXPRESSIONS

inclusion-exclusion principle　包含容斥原理

## EXERCISES 3.3

1. How many elements are in $A \cup B$ if there are 20 elements in $A$, 10 elements in $B$, and
   $A \cap B = \emptyset$?　　　$|A \cap B| = 1$?
   $|A \cap B| = 6$?　　　$A \subseteq B$?

2. Find the number of elements in $A \cup B \cup C$ if there are 100 elements in each set if
   a. the sets are pair wise disjoint;
   b. there are 50 common elements in each pair of sets and no elements in all three sets;
   c. there are 50 common elements in pair of sets and 25 elements in all three sets;
   d. the sets are equal.

3. Find the number of positive integers not exceeding 100 that are not divisible by 5 or by 7.

4. In a survey of 300 college students, it is found that 70 like Brussels sprouts, 80 like broccoli, 60 like cauliflower, 50 like both Brussels sprouts and broccoli, 40 like both Brussels sprouts and cauliflower, 30 like both broccoli and cauliflower, and 20 like all three vegetables. How many of the 300 students do not like any of these vegetables?

5. In a school there are 500, 300, 310 and 350 students in the courses of calculus, discrete mathematics, data structures, and programming languages, respectively. Among these students 15 are in both calculus and data structures, 220 are in both

calculus and programming languages, 200 are in both discrete mathematics and data structures, and 50 are in both discrete mathematics and programming languages. No student may take calculus and discrete mathematics or data structures and programming languages, concurrently. How many students are enrolled in a course of either calculus, discrete mathematics, data structures, or programming languages?

6. Suppose $A$, $B$, and $C$ represent the sets of bus stops of three bus routes through a suburb of your favorite city. Suppose $A$ has 25 stops, $B$ has 30 stops, and $C$ has 40 stops. Suppose further that $A$ and $B$ share 6 stops, $A$ and $C$ share 5 stops, and $B$ and $C$ share 4 stops. Lastly, suppose that $A$, $B$, and $C$ share 2 stops. Answer each of the following questions:

   a. How many distinct stops are on the three bus routes?
   b. How many stops for $A$ are not stops for $B$?
   c. How many stops for $A$ are not stops for both $B$ and $C$?
   d. How many stops for $A$ are not stops for any other bus route?

# 4

# Relations

Relationships between different entities exist everywhere in the real world, and it is also common to find relationships between objects in mathematics. Examples of such relationships include integers and their divisors, real numbers and their logarithms. If we go a little further, we can find that functions are used in mathematics to describe relationships between variables, and in set theory, equivalence relations are used to describe similarity among elements of sets. This chapter includes basic theory of relations and their elementary applications. It is easy to see that there are strong links between relations and sets.

## 4.1 Cartesian Products and Relations

The order of elements in a collection is often important. Since sets are unordered, a different structure is needed to represent ordered collection. This is provided by ordered $n$-tuple.

**DEFINITION 4.1.1** The **ordered $n$-tuple** $(a_1, a_2, \cdots, a_n)$ is the ordered collection that has $a_1$ as its first element, $a_2$ as its second element, $\cdots$, and $a_n$ as its $n$th element.

Two ordered $n$-tuples $(a_1, a_2, \cdots, a_n)$ and $(b_1, b_2, \cdots, b_n)$ are said to be equal if $a_i = b_i$ for $1 \leqslant i \leqslant n$, and we denote this by $(a_1, a_2, \cdots, a_n) = (b_1, b_2, \cdots, b_n)$. A 2-tuple is often called an **ordered pair**, and a 3-tuple may be called an **ordered triple**. Other words used in place of the word *tuple* are *vector* and *sequence*. Many of the discrete structures we will study in later chapters are based on the notion of the Cartesian product of sets, named after René Descartes (1596—1650).

**DEFINITION 4.1.2** Let $A$ and $B$ be sets. The **Cartesian product**, or **cross product** of $A$ and $B$, denoted by $A \times B$, is the set of all ordered pairs $(a,b)$ where $a \in A$ and $b \in B$. Hence
$$A \times B = \{(a,b) \mid a \in A \wedge b \in B\}$$

**EXAMPLE 4.1.1** Let $A = \{2, 3, 4\}$, $B = \{a, b\}$. Then
$A \times B = \{(2,a),(2,b),(3,a),(3,b),(4,a),(4,b)\}$
$B \times A = \{(a,2),(b,2),(a,3),(b,3),(a,4),(b,4)\}$
$A \times A = \{(2,2),(2,3),(2,4),(3,2),(3,3),(3,4),(4,2),(4,3),(4,4)\}$
$B \times B = \{(a,a),(a,b),(b,a),(b,b)\}$

**EXAMPLE 4.1.2** Let $A = \varnothing$ and $B = \{0,1\}$, what is $A \times B$? If we apply the definition of product, we must conclude that there are not any ordered pairs with first elements from the empty set. Therefore $A \times B = \varnothing$. It is easy to see that $A \times B$ is nonempty if and only if both $A$ and $B$ are nonempty sets.

The Cartesian product, or cross product, $A \times B$ and $B \times A$ are not equal (verified by EXAMPLE 4.1.1), unless $A = \varnothing$ or $B = \varnothing$ (so that $A \times B = B \times A = \varnothing$) or unless $A = B$. We can extend the definition of the Cartesian product to more than two sets.

**DEFINITION 4.1.3** The Cartesian product of sets $A_1, A_2, \cdots, A_n$, denoted by $A_1 \times A_2 \times \cdots \times A_n$, is $\{(a_1, a_2, \cdots, a_n) \mid a_i \in A_i, 1 \leq i \leq n\}$.

**EXAMPLE 4.1.3** What is the Cartesian product $A \times B \times C$, where $A = \{1,2\}$, $B = \{2,3\}$ and $C = \{a,b\}$?
*Solution*: $A \times B \times C = \{(1,2,a),(1,2,b),(1,3,a),(1,3,b),(2,2,a),(2,2,b),(2,3,a),(2,3,b)\}$.

The Cartesian product and the binary operations of union and intersection are interrelated in the following theorem:

**THEOREM 4.1.1** For any sets $A$, $B$, and $C$
1. $A \times (B \cap C) = (A \times B) \cap (A \times C)$
2. $A \times (B \cup C) = (A \times B) \cup (A \times C)$
3. $(A \cap B) \times C = (A \times C) \cap (B \times C)$
4. $(A \cup B) \times C = (A \times C) \cup (B \times C)$

*Proof*: We only prove (1) and (4) and leave the other parts to the readers. We use the same concept of set equality even though the elements here are ordered pairs.

1. For any $(a,b) \in A \times (B \cap C) \equiv a \in A \wedge b \in (B \cap C)$
$$\equiv a \in A \wedge (b \in B \wedge b \in C)$$
$$\equiv (a \in A \wedge b \in B) \wedge (a \in A \wedge b \in C)$$
$$\equiv (a,b) \in A \times B \wedge (a,b) \in A \times C$$
$$\equiv (a,b) \in (A \times B \cap A \times C)$$

4. For any $(a,b) \in (A \cup B) \times C \equiv (a \in A \cup B) \wedge b \in C$
$$\equiv (a \in A \wedge b \in C) \vee (a \in B \wedge b \in C)$$
$$\equiv (a,b) \in A \times C \vee (a,b) \in B \times C$$
$$\equiv (a,b) \in (A \times C \cup B \times C)$$

Returning to the Cartesian product of two sets, we shall find the subsets of this structure of great interest.

**DEFINITION 4.1.4** Let $A$ and $B$ be sets. A binary relation from $A$ to $B$ is a subset of $A \times B$. Any subset of $A \times A$ is called a **binary relation** on $A$.

In other words, a binary relation from $A$ to $B$ is a set $R$ of ordered pairs where the first element of each ordered pair comes from $A$ and the second element comes from $B$. We use the notation $aRb$ to denote that $(a,b) \in R$ and $a\not{R}b$ to denote that $(a,b) \notin R$. Moreover, when $(a,b)$ belongs to $R$, $a$ is said to be related to $b$ by $R$.

Binary relations represent relationships between the elements of two sets. We will omit the word binary when there is no danger of confusion.

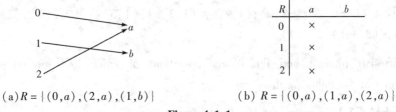

(a) $R = \{(0,a),(2,a),(1,b)\}$      (b) $R = \{(0,a),(1,a),(2,a)\}$

Figure 4.1.1

**EXAMPLE 4.1.4** Let $A = \{0,1,2\}$ and $B = \{a,b\}$. Then $\{(0,a),(0,b)\}$, $A \times B$, $\varnothing$, $\{(0,a),(2,a),(1,b)\}$ are relations from $A$ to $B$. Relations can be represented graphically, as shown in Figure 4.1.1 (a) using arrows to represent ordered pairs. Another way to represent relations is to use a table, which is done in Figure 4.1.1 (b). We will discuss representations of relations in more detail in Section 4.3.

**EXAMPLE 4.1.5** Let $A$ be the set $\{1, 2, 3, 4\}$. Which ordered pairs are in the relation $R = \{(a,b) \mid a \text{ divides } b\}$?

*Solution*: Since $(a,b)$ is in $R$ if and only if $a$ and $b$ are positive integers not exceeding 4 such that $a$ divides $b$, we can get that $R = \{(1,1), (1,2), (1,3), (1,4), (2,2), (2,4), (3,3), (4,4)\}$. The pairs in this relation are displayed both graphically and in tabular form in Figure 4.1.2.

**Figure 4.1.2** Graphical and tabular display of $R$

For finite sets $A$ and $B$ with $|A| = m$ and $|B| = n$, there are $2^{mn}$ relations from $A$ to $B$, (there are $2^{m^2}$ relations on $A$), including the empty relation as well as the relation $A \times B$ ($A \times A$) itself.

Since relations from $A$ to $B$ are subsets of $A \times B$, two relations from $A$ to $B$ can be combined in any way two sets can be combined.

**EXAMPLE 4.1.6** Let $A = \{1,2,3\}$, $B = \{a,b,c\}$. The relations $R_1 = \{(1,a), (2,b), (3,c)\}$ and $R_2 = \{(1,a), (1,b), (1,c), (3,c)\}$ can be combined to obtain

$R_1 \cup R_2 = \{(1,a), (1,b), (1,c), (2,b), (3,c)\}$
$R_1 \cap R_2 = \{(1,a), (3,c)\}$
$R_1 - R_2 = \{(2,b)\}$
$R_2 - R_1 = \{(1,b), (1,c)\}$
$R_1 \Delta R_2 = \{(1,b), (1,c), (2,b)\}$

**EXAMPLE 4.1.7** Let $R_1$ be the "Less than" relation on the set of real numbers and $R_2$ the "greater than" relation on the set of real numbers, that is, $R_1 = \{(a,b) \mid a < b\}$ and $R_2 = \{(a,b) \mid a > b\}$. What are $R_1 \cup R_2$, $R_1 \cap R_2$, $R_1 - R_2$, $R_2 - R_1$, and $R_1 \Delta R_2$?

*Solution*: We note that $(a,b) \in R_1 \cup R_2$ if and only if $(a,b) \in R_1$, or $(a,b) \in R_2$. Hence $(a,b) \in R_1 \cup R_2$ if and only if $a < b$ or $a > b$. Because the condition $a < b$ or $a > b$ is the same as the condition $a \neq b$, it follows that $R_1 \cup R_2 = \{(a,b) \mid a \neq b\}$. Meanwhile it is impossible for a pair $(a,b)$ to belong to both $R_1$ and $R_2$ since it is impossible for $a < b$ and $a > b$. It means that $R_1 \cap R_2 = \emptyset$. We also see that $R_1 - R_2 = R_1$, $R_2 - R_1 = R_2$, $R_1 \Delta R_2 = \{(a,b) \mid a \neq b\}$.

## WORDS AND EXPRESSIONS

| | |
|---|---|
| ordered pair | 序偶,有序对 |
| vector | 向量 |
| Cartesian product | 笛卡尔乘积 |
| cross product | 叉乘 |
| binary relation | 二元关系 |

## EXERCISES 4.1

1. List the ordered pairs in relation $R$ from $A = \{0,2,3,4\}$ to $B = \{0,1,2,3\}$ where $(a,b) \in R$ if and only if
    a. $a = b$
    b. $a + b = 4$
    c. $a > b$
    d. $a \mid b$
    e. $\gcd(a,b) = 1$
    f. $\text{lcm}(a,b) = 2$

2. Let $A = \{1,2,4,6,8,16\}$ and $B = \{1,2,3,4,5,6,7\}$. If $(x-2,5), (4, y-2) \in A \times B$. Can $(x-2,5) = (4, y-2)$?

3. Let $R_1 = \{(1,2), (2,3), (3,4)\}$ and $R_2 = \{(1,1), (1,2), (2,1), (2,2), (2,3), (3,1), (3,2), (3,3), (3,4)\}$ be relations from $\{1,2,3\}$ to $\{1,2,3,4\}$. Find
$R_1 \cup R_2 \qquad R_1 \cap R_2 \qquad R_1 - R_2 \qquad R_2 - R_1 \qquad R_1 \Delta R_2$

4. Assume that $A = \{1,2,3,4\}$. We define two relations $R_1 = \{(a,b) \mid \frac{a-b}{2}$ is an integer$\}$ and $R_2 = \{(a,b) \mid \frac{a-b}{3}$ is a positive integer$\}$ on set $A$. Find

$R_1 \cup R_2$     $R_1 \cap R_2$     $R_1 - R_2$     $R_2 - R_1$     $R_1 \Delta R_2$

5. Let $R_1 = \{(a,b) \mid a \equiv b (\bmod 3)\}$ and $R_2 = \{(a,b) \mid a \equiv b (\bmod 4)\}$. Find
$R_1 \cup R_2$     $R_1 \cap R_2$     $R_1 - R_2$     $R_2 - R_1$     $R_1 \Delta R_2$

6. Prove that
   a. $(A - B) \times C = A \times C - B \times C$
   b. $A \times B \subseteq C \times D$ if and only if $A \subseteq C$ and $B \subseteq D$

## 4.2 Properties of Relations

There are several properties that are useful to classify relations on a set. We will only introduce the most important ones.

**DEFINITION 4.2.1** A relation $R$ on a set $A$ is called **reflexive** if for all $a \in A$, $(a,a) \in R$. A relation $R$ on a set $A$ is called **irreflexive** if $(a,a) \notin R$ for each element $a \in A$.

**EXAMPLE 4.2.1** For $A = \{1, 2, 3, 4\}$, a relation $R \subseteq A \times A$ will be reflexive if and only if
$\{(1,1),(2,2),(3,3),(4,4)\} \subseteq R$. A relation $R \subseteq A \times A$ will be irreflexive if and only if $\{(1,1),(2,2),(3,3),(4,4)\} \cap R = \emptyset$. Consequently,
$R_1 = \{(1,1),(1,2),(2,2),(2,1),(3,4),(4,4),(3,3)\}$ is reflexive;
$R_2 = \{(1,1),(2,2),(3,4),(4,3)\}$ is neither reflexive nor irreflexive;
$R_3 = \{(1,2),(2,1),(3,4),(4,3)\}$ is irreflexive.

The set $I_A = \{(a,a) \mid a \in A\}$ is called the **diagonal relation** on $A$. From DEFINITION 4.2.1, a relation $R$ on $A$ is reflexive if and only if $I_A \subseteq R$, and $R$ is irreflexive if and only if $I_A \cap R = \emptyset$.

**EXAMPLE 4.2.2** Consider these relations on the set of integers:
$R_1 = \{(a,b) \mid a \leq b\}$,
$R_2 = \{(a,b) \mid a > b\}$,
$R_3 = \{(a,b) \mid a = b, \text{or } a = -b\}$,
$R_4 = \{(a,b) \mid a = b\}$,
$R_5 = \{(a,b) \mid a = b + 1\}$,
$R_6 = \{(a,b) \mid a + b \leq 3\}$.

Which of these relations are reflexive? Which of them are irreflexive?

*Solution*: $R_1$ is reflexive, since $a \leqslant a$ for every integer $a$. $R_2$ is irreflexive. $R_3$ and $R_4$ are reflexive. $R_5$ is irreflexive. $R_6$ is neither reflexive nor irreflexive.

**EXAMPLE 4.2.3** Is the "divides" relation on the set of positive integers reflexive, or irreflexive?

*Solution*: Since $a \mid a$ whenever $a$ is a positive integer, the "divides" relation is reflexive.

**EXAMPLE 4.2.4** Given a finite set $A$ with $|A| = n$, we have $|A \times A| = n^2$. So there are $2^{n^2}$ relations on $A$. How many of these are reflexive?

*Solution*: If $A = \{a_1, a_2, \cdots, a_n\}$, a relation $R$ on $A$ is reflexive if and only if $I_A = \{(a_i, a_i) \mid a_i \in A, 1 \leqslant i \leqslant n\} \subseteq R$. Considering the other $n^2 - n$ ordered pairs in $A \times A$ (those of the form $(a_i, a_j)$, where $i \neq j$ for $1 \leqslant i, j \leqslant n$) as we construct a reflexive $R$ on $A$, we either include or exclude each of these ordered pairs. So there are $2^{n^2-n}$ reflexive relations on $A$.

**DEFINITION 4.2.2** A relation $R$ on a set $A$ is called **symmetric** if $(b, a) \in R$ whenever $(a, b) \in R$, for all $a, b \in A$. A relation $R$ on a set $A$ is called **antisymmetric** if for all $a, b \in A$, $(a, b) \in R$ and $(b, a) \in R$ only if $a = b$.

Remark: Using quantifiers, we see that a relation $R$ on a set $A$ is symmetric if $\forall a \; \forall b \; ((a, b) \in R \rightarrow (b, a) \in R)$. Similarly, a relation $R$ on a set $A$ is antisymmetric if $\forall a \; \forall b \; ((a, b) \in R \land (b, a) \in R \rightarrow (a = b))$. In other words, a relation is antisymmetric if and only if $\forall a \; \forall b \; ((a \neq b) \rightarrow (a, b) \notin R \lor (b, a) \notin R)$. The terms *symmetric* and *antisymmetric* are not opposite, since a relation may have both of these properties or may lack both of them.

**EXAMPLE 4.2.5** Which of the relations from EXAMPLE 4.2.2 are symmetric and which are antisymmetric?

*Solution*: $R_3$, $R_4$ and $R_6$ are symmetric, while $R_1$, $R_2$, $R_4$ and $R_5$ are antisymmetric.

**EXAMPLE 4.2.6** With $A = \{1, 2, 3\}$, we have that

$R_1 = \{(1, 2), (2, 1), (2, 3), (3, 2), (1, 1)\}$ is symmetric, but not antisymmetric;

$R_2 = \{(1, 1), (2, 2), (3, 3), (2, 3)\}$ is reflexive and antisymmetric, but not symmetric;

$R_3 = \{(1, 1), (2, 2), (3, 3)\}$ is symmetric and antisymmetric;

$R_4 = \{(1,2),(2,1),(2,3),(3,3)\}$ is neither symmetric nor antisymmetric.

**EXAMPLE 4.2.7** Is the "divides" relation on the set of positive integers symmetric? Is it antisymmetric?

*Solution*: This relation is not symmetric since $1|2$, but $2|1$ is not true. It is antisymmetric, since if $a$ and $b$ are positive integers such that $a|b$ and $b|a$ then $a = b$.

To count the symmetric relations on $A = \{a_1, a_2, \cdots, a_n\}$, we write $A \times A$ as $A_1 \cup A_2$, where $A_1 = \{(a_i, a_i) | 1 \leq i \leq n\}$ and $A_2 = \{(a_i, a_j) | 1 \leq i,j \leq n, i \neq j\}$, so that every ordered pair in $A \times A$ is in exactly one of $A_1$ and $A_2$. We have that $|A_2| = |A \times A| - |A_1| = n^2 - n$, which is an even integer. The set $A_2$ contains $\dfrac{n^2 - n}{2}$ subsets $S_{ij}$ of the form $\{(a_i, a_j),(a_j, a_i)\}$ where $1 \leq i < j \leq n$. In constructing a symmetric relation $R$ on $A$, for each ordered pair in $A_1$, we have our usual choice of exclusion or inclusion. For each of the $\dfrac{n^2-n}{2}$ subsets $S_{ij}$ ($1 \leq i < j \leq n$) in $A_2$ we have the same two choices. So there are $2^n \cdot 2^{\frac{n^2-n}{2}} = 2^{\frac{n^2+n}{2}}$ symmetric relations on $A$.

**DEFINITION 4.2.3** Let $R$ be a relation from set $A$ to set $B$. The **inverse** relation of $R$ is a relation from $B$ to $A$, denoted by $R^{-1}$, and defined by the set of ordered pairs $\{(b, a) | (a, b) \in R\}$.

**EXAMPLE 4.2.8** Show that $R$ is a symmetric relation on a set $A$ if and only if $R = R^{-1}$.

*Proof*: Assume that $R$ is a symmetric relation on a set $A$. For any $(a, b) \in R$, we have that $(b, a) \in R$. But $(b, a) \in R^{-1}$. It follows that $R \subseteq R^{-1}$. On the other hand, if $(a, b) \in R^{-1}$, then $(b, a) \in R$. Thus $(a, b) \in R$ because $R$ is symmetric. It follows that $R^{-1} \subseteq R$. Therefore $R = R^{-1}$.

Conversely, let $R = R^{-1}$. If $(a, b) \in R$, then $(b, a) \in R^{-1} = R$. It means that $R$ is symmetric.

**DEFINITION 4.2.4** A relation $R$ on a set $A$ is called **transitive** if whenever $(a,b) \in R$ and $(b,c) \in R$, then $(a,c) \in R$ for all $a,b,c \in A$.

Remark: Using quantifiers we see that a relation $R$ on a set $A$ is transitive if $\forall a\ \forall b\ \forall c\ ((a,b) \in R \wedge (b,c) \in R \rightarrow (a,c) \in R)$.

**EXAMPLE 4.2.9** Is the "divides" relation on the set of positive integers transitive?

*Solution*: Suppose that $a$ divides $b$, and $b$ divides $c$. Then there are positive integers $k$ and $l$ such that $b = ak$ and $c = bl$. Hence, $c = a(kl)$, i.e., $a$ divides $c$. This relation is transitive.

**EXAMPLE 4.2.10** Which of the relations in EXAMPLE 4.2.2 are transitive?

*Solution*: $R_1$ is transitive since $a \leqslant b$ and $b \leqslant c$ imply $a \leqslant c$. $R_2$ is transitive since $a > b$ and $b > c$ imply $a > c$. $R_3$ is transitive since $a = \pm b$ and $b = \pm c$ imply $a = \pm c$. $R_4$ is clearly transitive. $R_5$ is not transitive since $(2,1)$ and $(1,0)$ belong to $R_5$, but $(2,0)$ does not. $R_6$ is not transitive since $(2,1)$ and $(1,2)$ belong to $R_6$, but $(2,2)$ does not.

**DEFINITION 4.2.5** Let $R$ be a relation from set $A$ to set $B$ and $S$ be a relation from $B$ to set $C$. The **composite** of $R$ and $S$ is the relation consisting of ordered pairs $(a,c)$, where $a \in A$, $c \in C$, and there exists an element $b \in B$ such that $(a,b) \in R$ and $(b,c) \in S$. We denote the composite of $R$ and $S$ by $R \circ S$:

$$R \circ S = \{(a,c) \mid a \in A \wedge c \in C \wedge \exists b (b \in B \wedge (a,b) \in R \wedge (b,c) \in S)\}$$

**EXAMPLE 4.2.11** Suppose $R$ is a relation from $\{1,2,3\}$ to $\{1,2,3,4\}$ defined by $R = \{(1,1),(1,4),(2,3),(3,1),(3,4)\}$. $S$ is a relation from $\{1,2,3,4\}$ to $\{0,1,2\}$ defined by $S = \{(1,0),(2,0),(3,1),(3,2),(4,1)\}$. What is the composite of the relations $R$ and $S$?

*Solution*: $R \circ S$ is constructed using all ordered pairs in $R$ and $S$, where the second element of the ordered pair in $R$ agrees with the first element of the ordered pair in $S$. For example, the ordered pairs $(2,3)$ in $R$ and $(3,1)$ in $S$ produce the ordered pair $(2,1)$ in $R \circ S$. Computing all ordered pairs in the composite, we find that $R \circ S = \{(1,0),(1,1),(2,1),(2,2),(3,0),(3,1)\}$.

There are some fundamental properties of composite relations. Let $R$, $S$ and $T$ be relations. Then:

1. $R \circ (S \circ T) = (R \circ S) \circ T$
2. $R \circ (S \cup T) = R \circ S \cup R \circ T$
3. $R \circ (S \cap T) \subseteq R \circ S \cap R \circ T$

We leave the proof of these properties to the readers.

**EXAMPLE 4.2.12** Let $R = \{(1,2),(3,4),(2,2)\}$, $S = \{(4,2),(2,5),(3,1),$

$(1,3)\}$. Find $R \circ S$, $S \circ R$, $R \circ (S \circ R)$, $(R \circ S) \circ R$, $R \circ R$, $S \circ S$, and $R \circ R \circ R$.

Solution: $R \circ S = \{(1,5),(3,2),(2,5)\}$
$S \circ R = \{(4,2),(3,2),(1,4)\} \; (\neq R \circ S)$
$(R \circ S) \circ R = \{(3,2)\}$
$R \circ (S \circ R) = \{(3,2)\}$
$R \circ R = \{(1,2),(2,2)\}$
$S \circ S = \{(4,5),(3,3),(1,1)\}$
$R \circ R \circ R = \{(1,2),(2,2)\}$

The powers of a relation $R$ can be recursively defined from the definition of composite of two relations.

**DEFINITION 4.2.6** Let $R$ be a relation on the set $A$. The powers $R^n$, $n = 1,2,3,4,\cdots$ are defined recursively by $R^1 = R$ and $R^{n+1} = R^n \circ R$.

**EXAMPLE 4.2.13** Let $R = \{(1,1),(2,1),(3,2),(4,3)\}$. Find the powers $R^n$, $n = 2,3,4,\cdots$.

Solution: $R^2 = R \circ R = \{(1,1),(2,1),(3,1),(4,2)\}$
$R^3 = R^2 \circ R = \{(1,1),(2,1),(3,1),(4,1)\}$
$R^4 = R^3 \circ R = \{(1,1),(2,1),(3,1),(4,1)\}$

$R^4$ is the same as $R^3$. It also follows that $R^n = R^3$ for $n = 5,6,7,\cdots$

The following theorem shows that the powers of a transitive relation are subsets of this relation.

**THEOREM 4.2.1** The relation $R$ on a set $A$ is transitive if and only if $R^n \subseteq R$ for $n = 1,2,3,\cdots$

Proof: We first prove the "if" part of the theorem. Suppose that $R^n \subseteq R$ for $n = 1,2,3,\cdots$. In particular $R^2 \subseteq R$. To see that this implies $R$ is transitive, note that if $(a,b) \in R$ and $(b,c) \in R$, then $(a,c) \in R^2$. Since $R^2 \subseteq R$, this means that $(a,c) \in R$. Hence $R$ is transitive.

We will use mathematical induction to prove the "only if" part of the theorem. Note that this part of the theorem is trivially true for $n = 1$.

Assume that $R^n \subseteq R$ where $n$ is a positive integer. This is the inductive hypothesis. To complete the inductive step we must show that this implies that $R^{n+1}$ is also a subset of

$R$. To show this, assume that $(a,b) \in R^{n+1}$. Then, since $R^{n+1} = R^n \circ R$, there is an element $x \in A$ such that $(a,x) \in R^n$ and $(x,b) \in R$. The inductive hypothesis implies that $(a,x) \in R$. Furthermore, since $R$ is transitive, and $(a,x) \in R$ and $(x,b) \in R$, it follows that $(a,b) \in R$. This shows that $R^{n+1} \subseteq R$, completing the proof.

## WORDS AND EXPRESSIONS

| | |
|---|---|
| reflexive | 自反的 |
| irreflexive | 反自反的 |
| symmetric | 对称的 |
| antisymmetric | 反对称的 |
| transitive | 传递的 |
| composite | 合成,复合 |

## EXERCISES 4.2

1. For each of the following relations on the set $\{1,2,3,4\}$, determine whether it is reflexive, irreflexive, symmetric, antisymmetric, or transitive.
   a. $\{(2,2),(2,3),(2,4),(3,2),(3,3),(3,4)\}$
   b. $\{(1,1),(1,2),(2,1),(2,2),(3,3),(4,4)\}$
   c. $\{(2,4),(4,2)\}$
   d. $\{(1,2),(2,3),(3,4)\}$
   e. $\{(1,1),(2,2),(3,3),(4,4)\}$
   f. $\{(1,3),(1,4),(2,3),(2,4),(3,1),(3,4)\}$

2. Determine whether the relation $R$ on the set of all integers is reflexive, irreflexive, symmetric, antisymmetric, or transitive, where $(x,y) \in R$ if and only if
   a. $x \neq y$
   b. $xy \geq 1$
   c. $x = y+1$ or $x = y-1$
   d. $x \equiv y \pmod 1$
   e. $x$ is a multiple of $y$
   f. $x$ and $y$ are both negative or both nonnegative
   g. $x = y^2$

3. Let $R = \{(1,2),(1,3),(2,4),(3,1)\}$ and $S = \{(2,1),(3,1),(3,2),(4,2)\}$

be relations. Find

$S \circ R$     $R \circ S$     $R^2$     $S^3$

4. For each of the following properties, show that if $R$ has the property, so does $R^n$.
   Reflexive     symmetric     transitive

5. Prove each of the following statements about binary relations.
   a. $R \circ (S \circ T) = (R \circ S) \circ T$
   b. $R \circ (S \cup T) = R \circ S \cup R \circ T$
   c. $R \circ (S \cap T) \subseteq R \circ S \cap R \circ T$

6. How many relations are there on a set with $n$ elements having the following properties?
   Irreflexive     antisymmetric     reflexive and symmetric

## 4.3 Representing Relations

There are many ways to represent a relation between finite sets. As we have seen, one way is to list its ordered pairs. In this section we will discuss two alternative methods for representing relations. One method uses zero-one matrices. The other method uses directed graphs.

Generally, matrices are appropriate for the representation of relations in computer programs. On the other hand, people often find the representation of relations using directed graphs useful for understanding the properties of the relations.

**DEFINITION 4.3.1** Suppose that $R$ is a relation from $A = \{a_1, a_2, \cdots, a_m\}$ to $B = \{b_1, b_2, \cdots, b_n\}$. We define the **relation matrix** $M = [m_{ij}]_{m \times n}$ to represent relation $R$ as follows:

$$m_{ij} = \begin{cases} 1, & \text{if } (a_i, b_j) \in R \\ 0, & \text{if } (a_i, b_j) \notin R \end{cases}$$

Note that the elements of the sets $A$ and $B$ have been listed in a particular, but arbitrary, order. Furthermore, when $A = B$ we use the same ordering for $A$ and $B$.

DEFINITION 4.3.1 says that the zero-one matrix representing $R$ has 1 as its $(i,j)$ entry when $a_i$ is related to $b_j$; and 0 in this position if $a_i$ is not related to $b_j$. Such a representation depends on the orderings used for $A$ and $B$.

**EXAMPLE 4.3.1** Let $A = \{1,2,3,4\}$ and $B = \{a,b,c\}$, $R = \{(1,a),(2,a),(3,b),(3,c),(4,a),(4,c)\}$ is a relation from $A$ to $B$. What is the relation matrix $M_R$?

*Solution*:

$$M_R = \begin{pmatrix} 1 & 0 & 0 \\ 1 & 0 & 0 \\ 0 & 1 & 1 \\ 1 & 0 & 1 \end{pmatrix}$$

The 1s in $M_R$ mean that the pairs $(1,a),(2,a),(3,b),(3,c),(4,a)$ and $(4,c)$ belong to $R$, and 0s mean that all of other elements do not belong to $R$.

**EXAMPLE 4.3.2** Let $A = \{x,y,z\}$, and $B = \{4,5,6,7\}$. Which ordered pairs are in the relation $R$ represented by the following matrix?

$$M_R = \begin{pmatrix} 0 & 1 & 0 & 0 \\ 1 & 0 & 0 & 1 \\ 1 & 1 & 0 & 0 \end{pmatrix}$$

*Solution*: We obtain that $R = \{(x,5),(y,4),(y,7),(z,4),(z,5)\}$ from the definition of relation matrix.

The matrix of a relation on a set should be a square matrix, which can be used to determine whether the relation has certain properties. Recall that a relation $R$ on $A$ is reflexive if $(a,a) \in R$, for $a \in A$. Thus, if $A = \{a_1, a_2, \cdots, a_n\}$, $R$ is reflexive if and only if $(a_i, a_i) \in R$ for $i = 1,2, \cdots, n$. Hence, $R$ is reflexive if and only if $m_{ii} = 1$, for $i = 1,2, \cdots, n$. In other words, $R$ is reflexive if and only if all the elements on the main diagonal of $M_R$ are equal to 1.

A relation $R$ is irreflexive if and only if $(a_i, a_i) \notin R$ for all $i = 1, 2, \cdots, n$. Consequently, the matrix of an irreflexive relation has the property that $m_{ii} = 0$ for $i = 1, 2, \cdots, n$.

A relation $R$ is symmetric if $(a,b) \in R$ implies $(b,a) \in R$. Consequently, a relation $R$ on the set $A = \{a_1, a_2, \cdots, a_n\}$ is symmetric if and only if $(a_j, a_i) \in R$ whenever $(a_i, a_j) \in R$. In terms of the entries of $M_R$, $R$ is symmetric if and only if $m_{ji} = 1$ whenever $m_{ij} = 1$. This also means that $m_{ji} = 0$ whenever $m_{ij} = 0$. Consequently, $R$ is symmetric if and only if $m_{ij} = m_{ji}$ for all pairs of integers $i$ and $j$ where $i = 1, 2, \cdots, n$ and $j = 1, 2, \cdots, n$. Recalling the definition of the transpose of a matrix in *Linear Algebra*, we see that $R$ is symmetric if and only if $M_R = M_R^t$, i.e., $M_R$ is a symmetric matrix.

A relation $R$ is antisymmetric if and only if $a = b$ whenever $(a, b) \in R$ and $(b, a) \in R$. This implies that the matrix of an antisymmetric relation has the property that if $m_{ij} = 1$ for $i \neq j$, then $m_{ji} = 0$. Or, in other words, either $m_{ij} = 0$ or $m_{ji} = 0$ when $i \neq j$.

**EXAMPLE 4.3.3** Suppose that the relation $R$ on a set is represented by the relation matrix

$$M_R = \begin{pmatrix} 1 & 0 & 1 \\ 1 & 1 & 0 \\ 1 & 0 & 1 \end{pmatrix}.$$

Is $R$ reflexive, symmetric, and/or antisymmetric?

*Solution*: Since all the main diagonal elements of this matrix are equal to 1, $R$ is reflexive.

Because $m_{12} \neq m_{21}$, $R$ is not symmetric. Also, since $m_{13} = m_{31}$ so $R$ is not antisymmetric.

In order to investigate the operations of relation matrices, we introduce the Boolean operations $\vee$ and $\wedge$ operating on pairs of bits, defined by

$$a_1 \wedge a_2 = \begin{cases} 1, & \text{if } a_1 = a_2 = 1 \\ 0, & \text{otherwise} \end{cases} \qquad a_1 \vee a_2 = \begin{cases} 1, & \text{if } a_1 = 1 \text{ or } a_2 = 1 \\ 0, & \text{otherwise} \end{cases}$$

**DEFINITION 4.3.2** Let $A = [a_{ij}]$ and $B = [b_{ij}]$ be $m \times n$ zero-one matrices. Then the **join** of $A$ and $B$, denoted by $A \vee B$, is the zero-one matrix with $(i, j)$ th entry $a_{ij} \vee b_{ij}$. The **meet** of $A$ and $B$, denoted by $A \wedge B$, is the zero-one matrix with $(i, j)$ th entry $a_{ij} \wedge b_{ij}$.

**EXAMPLE 4.3.4** Find the join and the meet of $A$ and $B$ defined by

$$A = \begin{pmatrix} 1 & 0 & 1 \\ 1 & 1 & 0 \end{pmatrix}; \quad B = \begin{pmatrix} 0 & 0 & 1 \\ 1 & 0 & 1 \end{pmatrix}.$$

*Solution*: $A \vee B = \begin{pmatrix} 1 \vee 0 & 0 \vee 0 & 1 \vee 1 \\ 1 \vee 1 & 1 \vee 0 & 0 \vee 1 \end{pmatrix} = \begin{pmatrix} 1 & 0 & 1 \\ 1 & 1 & 1 \end{pmatrix};$

$A \wedge B = \begin{pmatrix} 1 \wedge 0 & 0 \wedge 0 & 1 \wedge 1 \\ 1 \wedge 1 & 1 \wedge 0 & 0 \wedge 1 \end{pmatrix} = \begin{pmatrix} 0 & 0 & 1 \\ 1 & 0 & 0 \end{pmatrix}.$

**DEFINITION 4.3.3** Let $A = [a_{ij}]$ be an $m \times k$ zero-one matrix and $B = [b_{ij}]$ a $k \times n$ zero-one matrix. Then the **Boolean product** of $A$ and $B$, denoted by $A \odot B$, is an $m \times n$ matrix with $(i, j)$ th entry $c_{ij}$, where $c_{ij} = (a_{i1} \wedge b_{1j}) \vee (a_{i2} \wedge b_{2j}) \vee \cdots \vee (a_{ik} \wedge b_{kj})$.

Note that the Boolean product of $A$ and $B$ is obtained in an analogous way to the ordinary product of these matrices, but with addition replaced by the operation $\vee$ and multiplication replaced by the operation $\wedge$.

**EXAMPLE 4.3.5** Find the Boolean product of $A$ and $B$, where

$$A = \begin{pmatrix} 1 & 1 & 0 \\ 0 & 0 & 1 \\ 1 & 0 & 1 \end{pmatrix}; \qquad B = \begin{pmatrix} 1 & 0 \\ 1 & 1 \\ 0 & 1 \end{pmatrix}.$$

*Solution:* The Boolean product $A \odot B$ is given by

$$A \odot B = \begin{pmatrix} (1 \wedge 1) \vee (1 \wedge 1) \vee (0 \wedge 0) & (1 \wedge 0) \vee (1 \wedge 1) \vee (0 \wedge 1) \\ (0 \wedge 1) \vee (0 \wedge 1) \vee (1 \wedge 0) & (0 \wedge 0) \vee (0 \wedge 1) \vee (1 \wedge 1) \\ (1 \wedge 1) \vee (0 \wedge 1) \vee (1 \wedge 0) & (1 \wedge 0) \vee (0 \wedge 1) \vee (1 \wedge 1) \end{pmatrix}$$

$$= \begin{pmatrix} 1 \vee 1 \vee 0 & 0 \vee 1 \vee 0 \\ 0 \vee 0 \vee 0 & 0 \vee 0 \vee 1 \\ 1 \vee 0 \vee 0 & 0 \vee 0 \vee 1 \end{pmatrix}$$

$$= \begin{pmatrix} 1 & 1 \\ 0 & 1 \\ 1 & 1 \end{pmatrix}.$$

Suppose that $R_1$ and $R_2$ are relations on a set $A$ represented by the matrices $M_{R_1}$ and $M_{R_2}$, respectively. The matrix representing the union of these relations has a 1 in the positions where either $M_{R_1}$ or $M_{R_2}$ has a 1. The matrix representing the intersection of these relations has a 1 in the positions where both $M_{R_1}$ and $M_{R_2}$ have a 1. Thus, the matrices representing the union and intersection of the relations are $M_{R_1 \cup R_2} = M_{R_1} \vee M_{R_2}$ and $M_{R_1 \cap R_2} = M_{R_1} \wedge M_{R_2}$.

**EXAMPLE 4.3.6** Suppose that the relations $R_1$ and $R_2$ on a set $A$ are represented by the matrices

$$M_{R_1} = \begin{pmatrix} 0 & 0 & 1 \\ 1 & 0 & 1 \\ 1 & 0 & 0 \end{pmatrix} \text{ and } M_{R_2} = \begin{pmatrix} 0 & 0 & 1 \\ 1 & 1 & 0 \\ 0 & 0 & 1 \end{pmatrix}.$$

What are the matrices representing $R_1 \cup R_2$ and $R_1 \cap R_2$?

Solution: $M_{R_1 \cup R_2} = M_{R_1} \vee M_{R_2} = \begin{pmatrix} 0 & 0 & 1 \\ 1 & 1 & 1 \\ 1 & 0 & 1 \end{pmatrix}$,

$M_{R_1 \cap R_2} = M_{R_1} \wedge M_{R_2} = \begin{pmatrix} 0 & 0 & 1 \\ 1 & 0 & 0 \\ 0 & 0 & 0 \end{pmatrix}$.

We now turn to determine the matrix for the composite of relations. Suppose that $R$ is a relation from set $A$ to $B$ and $S$ is a relation from $B$ to $C$. Suppose $A = \{a_1, a_2, \cdots, a_m\}$, $B = \{b_1, b_2, \cdots, b_n\}$ and $C = \{c_1, c_2, \cdots, c_p\}$. Let the zero-one matrices for $R \circ S$, $R$ and $S$ be $M_{R \circ S} = [t_{ij}]$, $M_R = [r_{ij}]$, and $M_S = [s_{ij}]$, respectively. An ordered pair belongs to $R \circ S$ if and only if there is an element $b_k \in B$ such that $(a_i, b_k)$ belongs to $R$ and $(b_k, c_j)$ belongs to $S$. It follows that $t_{ij} = 1$ if and only if $r_{ik} = s_{kj} = 1$ for some $k$. From the definition of the Boolean product, this means that $M_{R \circ S} = M_R \odot M_S$.

**EXAMPLE 4.3.7** Find the matrix representing the relation $R \circ S$, where the matrices representing $R$ and $S$ are

$$M_R = \begin{pmatrix} 0 & 0 & 1 \\ 1 & 1 & 0 \\ 1 & 0 & 1 \end{pmatrix} \text{ and } M_S = \begin{pmatrix} 0 & 1 & 0 \\ 0 & 0 & 1 \\ 1 & 1 & 0 \end{pmatrix}.$$

Solution: The relation matrix of $R \circ S$ is

$$M_{R \circ S} = M_R \odot M_S = \begin{pmatrix} 1 & 1 & 0 \\ 0 & 1 & 1 \\ 1 & 1 & 0 \end{pmatrix}.$$

**EXAMPLE 4.3.8** Find the relation matrix of $R^2$, where the relation matrix of $R$ is

$$M_R = \begin{pmatrix} 0 & 0 & 1 \\ 0 & 1 & 1 \\ 1 & 1 & 0 \end{pmatrix}.$$

Solution: The matrix for $R^2$ is

$$M_{R^2} = M_R \odot M_R = \begin{pmatrix} 1 & 1 & 0 \\ 1 & 1 & 1 \\ 0 & 1 & 1 \end{pmatrix}.$$

Pictorial representation is another important way to represent a relation, in which each element of the set is represented by a point, and each ordered pair is represented using an arc with its direction indicated by an arrow. We use pictorial representations when we think relations on a finite set as directed graphs.

**DEFINITION 4.3.4** A **directed graph**, or **digraph**, consists of a set $V$ of **vertices** (or **nodes**) together with a set $E$ of ordered pairs of elements of $V$ called **edges** (or **arcs**). The vertex $a$ is called the **initial vertex** of the edge $(a,b)$, and the vertex $b$ is called the **terminal vertex** of this edge. An edge of the form $(a,a)$ is represented using an arc from the vertex $a$ back to itself. Such an edge is called a **loop**.

A relation $R$ on a set $A$ is represented by the directed graph that has the elements of $A$ as its vertices and the ordered pairs $(a,b)$, where $(a,b) \in R$, as its edges. This assignment sets up a one-to-one correspondence between the relations on a set $A$ and the directed graphs with $A$ as their set of vertices. Directed graphs give a visual display of information about relations. They are often used to study relations and their properties.

**EXAMPLE 4.3.9** Let $R = \{(a,a),(a,b),(b,a),(b,c),(b,d),(c,d),(d,a)\}$. Then the directed graph of the relation $R$ on the set $\{a, b, c, d\}$ is depicted in Figure 4.3.1.

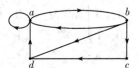

**Figure 4.3.1** Directed graph of $R$ in EXAMPLE 4.3.9

**EXAMPLE 4.3.10** What are the ordered pairs in the relation $R$ represented by the directed graph shown in Figure 4.3.2?

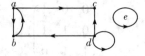

**Figure 4.3.2** Directed graph for EXAMPLE 4.3.10

*Solution*: The relation is $R = \{(a,b),(b,a),(a,c),(d,c),(d,b),(d,d),(e,e)\}$. Each of these pairs corresponds to an edge of the digraph, with $(d,d)$ and $(e,e)$ corresponding to loops.

The directed graph representing a relation can be used to determine whether the relation has certain properties. For instance, a relation is reflexive if and only if there is a loop at every vertex of the directed graph, so that every ordered pair of the form $(a,a)$ occurs in the relation. Similarly, a relation is irreflexive if and only if its digraph contains no loops. A relation is symmetric if and only if for every edge between distinct vertices in its digraph there is an edge with opposite direction, so that $(b,a)$ is also in the relation whenever $(a,b)$ is in the relation. Furthermore, a relation is antisymmetric if and only if there are never two edges in opposite directions between distinct vertices. Finally, a relation is transitive if and only if whenever there is an edge from a vertex $a$ to a vertex $b$ and an edge from $b$ to a vertex $c$, there is an edge from $a$ to $c$.

**EXAMPLE 4.3.11** Determine whether the relations represented by the digraphs shown in Figure 4.3.3 are reflexives, symmetric, antisymmetric, and/or transitive.

(a) Digraph for $R$          (b) Digraph for $S$

**Figure 4.3.3**

*Solution*: Since there is a loop at every vertex of the digraph for $R$, it is reflexive. Obviously $R$ is not irreflexive. Also, $R$ is neither symmetric since there is an edge from $a$ to $c$ but not one from $c$ to $a$, nor antisymmetric since there are edges in both directions connecting $a$ and $b$. Finally, $R$ is not transitive since there is an edge from $b$ to $a$ and an edge from $a$ to $d$, but no edge from $b$ to $d$.

Since no loop is presented at any vertex of the digraph of $S$, this relation is irreflexive. It is symmetric but not antisymmetric, since every edge between distinct vertices is accompanied by an edge in the opposite direction. $S$ is not transitive, since $(b,a)$ and $(a,c)$ belong to $S$, but $(b,c)$ does not belong to $S$.

## WORDS AND EXPRESSIONS

| | |
|---|---|
| matrix | 矩阵 |
| Boolean Product | 布尔积 |

directed graph (或 digraph)    有向图
vertex    点
edge    边
initial vertex    起点
terminal vertex    终点
loop    环

# EXERCISES 4.3

1. Represent each of the following relations on $\{1, 2, 3\}$ with a matrix (with the elements in increasing order).
   a. $\{(1, 1), (2, 3), (3, 3)\}$
   b. $\{(1, 2), (2, 1), (2, 2), (3, 2)\}$
   c. $\{(1, 1), (1, 3), (2, 2), (2, 3), (3, 3)\}$
   d. $\{(1, 3), (3, 1)\}$

2. List the ordered pairs in the relations on $\{1, 2, 3, 4\}$ corresponding to the following matrices (where the rows and columns correspond to the integers listed in increasing order).

$$\begin{pmatrix} 1 & 0 & 1 & 1 \\ 1 & 1 & 0 & 1 \\ 1 & 1 & 1 & 0 \\ 0 & 0 & 0 & 1 \end{pmatrix} \quad \begin{pmatrix} 0 & 1 & 1 & 1 \\ 1 & 0 & 1 & 0 \\ 0 & 0 & 1 & 0 \\ 1 & 0 & 0 & 1 \end{pmatrix}$$

3. Determine whether the relations represented by the matrices in Exercise 2 are reflexive, irreflexive, symmetric, antisymmetric, and/or transitive.

4. Suppose $R$ is a relation on a finite set $A$. How can the matrix for $R^{-1}$, the inverse of the relation $R$, be found from the relation matrix for $R$?

5. Let $R$ and $S$ be relations on a set $A$ represented by the matrices

$$M_R = \begin{pmatrix} 0 & 1 & 0 \\ 1 & 0 & 1 \\ 0 & 1 & 1 \end{pmatrix} \text{ and } M_S = \begin{pmatrix} 1 & 1 & 0 \\ 0 & 1 & 0 \\ 1 & 0 & 0 \end{pmatrix}.$$

Find the Matrices that represent the following relations:

$R \cup S$    $R \cap S$    $R \circ S$    $S \circ R$

$R \Delta S$      $R^2$      $R^3$      $R^4$

6. Draw the digraph representing each of the relations in Exercise 1.

7. List the ordered pairs in the relations represented by the following digraphs.

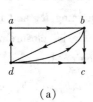   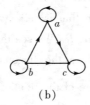      

(a)                (b)                (c)                (d)

8. Determine whether the relations represented by the digraphs shown in Exercise 7 are reflexive, irreflexive, symmetric, antisymmetric, and/or transitive.

## 4.4 Closure of Relations

We have seen how to construct a new relation by compositing two existing relations. Let's look at another way to construct a new relation from an existing relation. We will start with a binary relation $R$ and try to construct another relation containing $R$ that also satisfies some particular properties.

If $R$ is a relation on a set $A$ and $P$ is a property, such as being reflexive, symmetric, or transitive, then the $P$ **closure** of $R$ is the smallest binary relation that contains $R$ and satisfies property $P$. We denote the $P$ closure of $R$ by $P(R)$. If $R$ already satisfies property $P$, then we have $R = P(R)$.

We will be concerned with three properties: reflexivity, symmetry, and transitivity. The **reflexive closure** of $R$ is denoted by $r(R)$, the **symmetric closure** of $R$ is denoted by $s(R)$, and the **transitive closure** of $R$ is denoted by $t(R)$.

Our goal is to find some techniques to compute these closures. We will start with a running example to introduce the main idea.

The relation $R = \{(a,a),(a,b),(b,a),(b,c)\}$ on the set $A = \{a,b,c\}$ is not reflexive. We will compute the reflexive closure of $R$. This can be done by adding $(b,b)$ and $(c,c)$ to the relation since these two pairs are missing in $R$. Furthermore, any reflexive relation that contains $R$ must also contain $(b,b)$ and $(c,c)$. Thus $r(R) =$

$\{(a,a),(a,b),(b,a),(b,c),(b,b),(c,c)\}$, because this relation contains $R$, and also it is reflexive and is contained within every reflexive relation that contains $R$.

As shown in this example, given a relation $R$ on a set $A$, the reflexive closure of $R$ can be formed by adding to $R$ all pairs of the form $(x,x)$ for $x \in A$, which is not in $R$. The addition of these pairs produces a new relation that is reflexive, containing $R$, and is contained within any reflexive relation containing $R$. We see that the reflexive closure of $R$ is $r(R) = R \cup I_A$, where $I_A = \{(a,a) \mid a \in A\}$ is the diagonal relation on $A$.

**EXAMPLE 4.4.1** What is the reflexive closure of the relation $R = \{(a,b) \mid a > b\}$ on the set of integers?

*Solution*: The reflexive closure of $R$ is
$$R \cup I_Z = \{(a,b) \mid a > b\} \cup \{(a,a) \mid a \in \mathbf{Z}\} = \{(a,b) \mid a \geq b\}.$$

The relation $R = \{(a,a),(a,b),(b,a),(b,c)\}$ on $\{a,b,c\}$ is not symmetric. How can we construct a symmetric relation that is as small as possible and contains $R$? To do this, we need to add the pair $(c,b)$, since this is the only pair of the form $(y,x)$ not in $R$ while $(x,y) \in R$. After adding the pair the new relation is symmetric and contains $R$. Furthermore, *any* symmetric relation that contains $R$ must contain $(c,b)$. Consequently, the symmetric closure of $R$ is $s(R) = \{(a,a),(a,b),(b,a),(b,c),(c,b)\}$.

As illustrated by this example, the symmetric closure of a relation $R$ can be constructed by adding all ordered pairs of the form $(y,x)$ if $(x,y)$ is in the relation, but $(y,x)$ is not. Adding these pairs produces a relation that is symmetric, contains $R$, and is contained in any symmetric relation that contains $R$.

From the above discussion we see that the symmetric closure of $R$ is $s(R) = R \cup R^{-1}$ (where $R^{-1}$ is the inverse relation of $R$).

**EXAMPLE 4.4.2** What is the symmetric closure of $R = \{(a,b) \mid a > b\}$ on the set of positive integers?

*Solutions*: The symmetric closure of $R$ is
$$s(R) = R \cup R^{-1} = \{(a,b) \mid a > b\} \cup \{(b,a) \mid a > b\} = \{(a,b) \mid a \neq b\}.$$

If $R$ is not transitive, how can we produce a transitive closure? Can the transitive closure of a relation $R$ be produced by adding all the pairs of the form $(x,z)$ where $(x,y)$ and $(y,z)$ are already in the relation? Consider the relation $R = \{(1,3),$

$(1,4)$, $(2,1)$, $(3,2)\}$ on $\{1, 2, 3, 4\}$, which is not transitive since it does not contain all pairs of the form $(x,z)$ when $(x,y)$ and $(y,z)$ are in $R$. The pairs of this form missing in $R$ are $(1, 2)$, $(2, 3)$, $(2, 4)$, and $(3, 1)$. But adding these pairs does not produce a transitive relation, since the resulting relation contains $(3, 1)$ and $(1, 4)$ but $(3, 4)$ is missing. This shows that constructing the transitive closure of a relation is more complicated than constructing either the reflexive or the symmetric closure.

We will see that representing relations by directed graphs helps in the construction of transitive closure. We now introduce some terminology that we will use for this purpose.

**DEFINITION 4.4.1** A **directed walk** (abbreviated as **walk**) from $a$ to $b$ in a directed graph $G$ is a sequence of edges $(x_0, x_1)$, $(x_1, x_2)$, $(x_2, x_3)$, $\cdots$, $(x_{n-1}, x_n)$ in $G$, where $n$ is a nonnegative integer, $x_0 = a$ and $x_n = b$. This is a sequence of edges such that the terminal vertex of an edge is the same as the initial vertex of the next edge. This walk can be denoted by $x_0 x_1 x_2 x_3 \cdots x_n$ and has **length** $n$. A walk of length $n \geqslant 1$ that begins and ends at the same vertex is called a **closed walk**. A walk of a directed graph is called a **directed trail** (abbreviated as **trail**) if no edges in the walk occur more than once. A trail of a directed graph is called a **directed path** (abbreviated as **path**) if all vertices on the trail are distinct. A closed path is called a **circuit** or a **cycle**.

A walk (trail) in a directed graph can pass through a vertex more than once; moreover, an edge in a directed graph can occur more than once in a walk. However, any vertex cannot occur more than once in a path.

**EXAMPLE 4.4.3** Which of the following are walks, trails, or paths in the directed graph shown in Figure 4.4.1?

$abcdef$; $adade$; $cdeabc$; $dabcae$; $abcdada$; $adefda$

What are the lengths of those that are walks (trails, or paths)? Which of the walks are circuits?

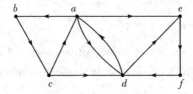

**Figure 4.4.1** The directed graph for EXAMPLE 4.4.3

*Solution*: Since $(a,b),(b,c),(c,d),(d,e)$ and $(e,f)$ are edges, *abcdef* is a path with length 5. Also, *adade* is a walk of length 4 since $(a,d),(d,a),(a,d)$ and $(d,e)$ are edges. The trail *dabcae* is of length 5. *abcdada* is a closed walk with length 6 since it begins and ends at the same vertex. *adefda* is a closed trail of length 5. However, *cdeabc* is not a walk because $(e,a)$ is not an edge.

The term *walk* also applies to relations. Carrying over the definition from directed graphs to relations, there is a walk from $a$ to $b$ in $R$ if there is a sequence of elements $a$, $x_1, x_2, x_3, \cdots, x_{n-1}, b$ where $(a, x_1) \in R$, $(x_1, x_2) \in R$, $\cdots$, and $(x_{n-1}, b) \in R$. Hence, we have the following theorem.

**THEOREM 4.4.1** Let $R$ be a relation on a set $A$. There is a walk of length $n$, where $n$ is a positive integer, from $a$ to $b$ if and only if $(a,b) \in R^n$.

*Proof*: We prove the theorem by induction. There is a walk from $a$ to $b$ of length one if and only if $(a,b) \in R$, so the theorem is true when $n = 1$.

Assume that the theorem is true for the positive integer $n$. This is the inductive hypothesis. There is a walk of length $n+1$ from $a$ to $b$ if and only if there is an element $c \in A$ such that there is a walk of length one from $a$ to $c$, so $(a,c) \in R$, and a walk of length $n$ from $c$ to $b$, that is, $(c,b) \in R^n$. Consequently, by the inductive hypothesis, there is a walk of length $n+1$ from $a$ to $b$ if and only if there is an element $c$ where $(a,c) \in R$ and $(c,b) \in R^n$. But there is such an element if and only if $(a,b) \in R^{n+1}$. Therefore, there is a walk of length $n+1$ from $a$ to $b$ if and only if $(a,b) \in R^{n+1}$. This finishes the proof.

We now show that finding the transitive closure of a relation is equivalent to determining which pairs of vertices in the associated directed graph are connected by a walk.

**THEOREM 4.4.2** The transitive closure of a relation $R$ is

$$t(R) = R \cup R^2 \cup R^3 \cup \cdots = \bigcup_{n=1}^{\infty} R^n.$$

*Proof*: For convenience, let $R^* = \bigcup_{n=1}^{\infty} R^n$.

Note that $R \subseteq R^*$. To show that $R^*$ is the transitive closure of $R$, we must also show that $R^*$ is transitive and that $R^* \subseteq S$ whenever $S$ is a transitive relation containing $R$.

If $(a,b) \in R^*$ and $(b,c) \in R^*$, then there are walks from $a$ to $b$ and from $b$ to $c$ in $R$. We obtain a walk from $a$ to $c$ by starting with the walk from $a$ to $b$ and following it with the walk from $b$ to $c$. Thus, $(a,c) \in R^*$, it follows that $R^*$ is transitive.

Suppose that $S$ is a transitive relation containing $R$. Since $S$ is transitive, $S^n$ is also transitive (by exercise 4 of Section 4.2) and $S^n \subseteq S$ (by THEOREM 4.2.1). It follows that $\bigcup_{n=1}^{\infty} S^n \subseteq S$. If $R \subseteq S$, then $R^* \subseteq \bigcup_{n=1}^{\infty} S^n$, because any walk in $R$ is also a walk in $S$. Consequently $R^* \subseteq S$. Hence, any transitive relation that contains $R$ must also contain $R^*$. Therefore $R^*$ is the transitive closure of $R$.

During producing the transitive closure, we don't need to examine arbitrary long walks to determine whether there is a walk between two vertices in a finite directed graph. As LEMMA 4.4.1 shows, it is sufficient to examine the walks containing no more than $n$ edges, where $n$ is the cardinality of the set.

**LEMMA 4.4.1** Let $A$ be a set with $n$ elements, and let $R$ be a relation on $A$. If there is a walk of length at least one in $R$ from $a$ to $b$, then there is a path with length not exceeding $n$. Moreover, when $a \neq b$, if there is a walk of length at least one in $R$ from $a$ to $b$, then there is such a path with length not exceeding $n-1$.

We leave the proof to readers as an exercise.

From LEMMA 4.4.1, we see that if $R$ is a relation on a finite set with $n$ elements, the transitive closure of $R$ is $t(R) = R \cup R^2 \cup \cdots \cup R^n$.

**EXAMPLE 4.4.4** Let $R = \{(a,b),(b,c),(c,d)\}$ be a relation on the set $A = \{a,b,c\}$, what is the transitive closure of $R$?
*Solution*: Since $R = \{(a,b),(b,c),(c,d)\}$, $R^2 = R \circ R = \{(a,c),(b,d)\}$, $R^3 = \{(a,d)\}$. Therefore, the transitive closure of $R$ is
$$t(R) = R \cup R^2 \cup R^3 = \{(a,b),(b,c),(c,d),(a,c),(b,d),(a,d)\}.$$

**EXAMPLE 4.4.5** Find the zero-one matrix of the transitive closure of relation $R$ where
$$M_R = \begin{pmatrix} 1 & 0 & 1 \\ 0 & 1 & 0 \\ 1 & 1 & 0 \end{pmatrix}.$$

*Solution*: The zero-one matrix of $t(R)$ is $M_{t(R)} = M_R \vee M_{R^2} \vee M_{R^3}$.

Since $M_R = \begin{pmatrix} 1 & 0 & 1 \\ 0 & 1 & 0 \\ 1 & 1 & 0 \end{pmatrix}$, we compute $M_{R^2}$ and $M_{R^3}$ as follows:

$$M_{R^2} = M_R \odot M_R = \begin{pmatrix} 1 & 1 & 1 \\ 0 & 1 & 0 \\ 1 & 1 & 1 \end{pmatrix} \text{ and } M_{R_3} = (M_R \odot M_R) \odot M_R = \begin{pmatrix} 1 & 1 & 1 \\ 0 & 1 & 0 \\ 1 & 1 & 1 \end{pmatrix}.$$

It follows that

$$M_{t(R)} = \begin{pmatrix} 1 & 0 & 1 \\ 0 & 1 & 0 \\ 1 & 1 & 0 \end{pmatrix} \vee \begin{pmatrix} 1 & 1 & 1 \\ 0 & 1 & 0 \\ 1 & 1 & 1 \end{pmatrix} \vee \begin{pmatrix} 1 & 1 & 1 \\ 0 & 1 & 0 \\ 1 & 1 & 1 \end{pmatrix} = \begin{pmatrix} 1 & 1 & 1 \\ 0 & 1 & 0 \\ 1 & 1 & 1 \end{pmatrix}.$$

Now, we list the construction techniques for all three closures.

If $R$ is a binary relation on a set $A$ with $|A| = n$, then

1. $r(R) = R \cup I_A$      ($I_A$ is the diagonal relation)
2. $s(R) = R \cup R^{-1}$      ($R^{-1}$ is the inverse relation of $R$)
3. $t(R) = R \cup R^2 \cup \cdots \cup R^n$

It follows that

4. $M_{r(R)} = M_R \vee M_{I_A}$
5. $M_{s(R)} = M_R \vee M_{R^{-1}}$
6. $M_{t(R)} = M_R \vee M_{R^2} \vee \cdots \vee M_{R^n}$

where $M_{R^m} = M_{R \circ R^{m-1}} = M_R \odot M_{R^{m-1}}$, $1 \leqslant m \leqslant n$.

The conclusion 6 above can be used as a basis of an algorithm for computing the matrix of the transitive closure. To find this matrix, Boolean powers of $M_R$, up to the $n$th power, are computed. As each power has been calculated, its join with the join of all smaller powers is formed. When this is done with the $n$th power, the matrix for $t(R)$ has been found. This procedure is displayed as the following algorithm.

**Algorithm 4.4.1** A procedure for computing the transitive closure (*transitive closure*) ($M_R$: zero-one $n \times n$ matrix)

    $A := M_R$
    $B := A$
    For $i := 2$ to $n$
    Begin
        $A := A \odot M_R$
        $B := B \vee A$
    End {$B$ is the Zero-one matrix for $t(R)$}.

We can easily find the number of bit operations used by Algorithm 4.4.1 to determine the transitive closure of a relation. Before computing the Boolean powers $M_R, M_{R^2}, \cdots, M_{R^n}$, we need to find $n-1$ Boolean products of $n \times n$ zero-one matrices. Each of these Boolean products involves $n^2(2n-1)$ bit operations. Hence, these products can be computed using $n^2(2n-1)(n-1)$ bit operations.

To find $M_{t(R)}$ from the $n$ Boolean powers of $M_R$, $n-1$ joins of zero-one matrices need to be found. Computing each of those joins uses $n^2$ bit operations. Hence, $(n-1)n^2$ bit operations are used in this part of the computation. Therefore, when Algorithm 4.4.1 is used, the matrix of the transitive closure of a relation on a set with $n$ elements can be found using $n^2(2n-1)(n-1) + n^2(n-1) = 2n^3(n-1)$ bit operations. There is a more efficient algorithm for finding transitive closure, called Warshall's algorithm, which can find the transitive closure using only $2n^3$ bit operations.

Suppose that $R$ is a relation on a set with $n$ elements $v_1, v_2, \cdots, v_n$. Warshall's algorithm is based on the construction of a sequence of zero-one matrices $W_0, W_1, \cdots, W_n$, where $W_0 = M_R$ is the zero-one matrix of the relation, and $W_k = [w_{ij}^{(k)}]$, where $w_{ij}^{(k)} = 1$ if there is a path from $v_i$ to $v_j$ such that all the internal vertices of this path are in the set $\{v_1, v_2, \cdots, v_k\}$ (the first $k$ vertices in the list) and is 0 otherwise. Note that $W_n = M_{t(R)}$, since the $(i,j)$th entry of $M_{t(R)}$ is 1 if and only if there is a path from $v_i$ to $v_j$ with all internal vertices in the set $\{v_1, v_2, \cdots, v_n\}$. The following example illustrates what the matrix $W_k$ represents.

**EXAMPLE 4.4.6** Let $R$ be the relation with directed graph shown in Figure 4.4.2. Let $a, b, c, d$ be a listing of the elements of the set. Find the matrices $W_0, W_1, W_2, W_3$ and $W_4$. Note that the matrix $W_4$ is the transitive closure of $R$.

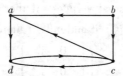

**Figure 4.4.2** Directed graph for $R$ in EXAMPLE 4.4.6

*Solution*: Let $v_1 = a, v_2 = b, v_3 = c, v_4 = d$. $W_0$ is the matrix of the relation. Hence,

$$W_0 = M_R = \begin{pmatrix} 0 & 0 & 0 & 1 \\ 1 & 0 & 1 & 0 \\ 1 & 0 & 0 & 1 \\ 0 & 0 & 1 & 0 \end{pmatrix}.$$

$W_1$ has 1 as its $(i,j)$th entry if there is a path from $v_i$ to $v_j$ that has only $v_1 = a$ as an internal vertex. Note that all paths of length one can still be used since they have no internal vertices. Also, there is now an allowable path from $b$ to $d$, namely, $b,a,d$. Hence,

$$W_1 = \begin{pmatrix} 0 & 0 & 0 & 1 \\ 1 & 0 & 1 & 1 \\ 1 & 0 & 0 & 1 \\ 0 & 0 & 1 & 0 \end{pmatrix}.$$

$W_2$ has 1 as its $(i,j)$th entry if there is a path from $v_i$ to $v_j$ that has only $v_1 = a$ and/or $v_2 = b$ as its internal vertices, if there is any. Since there are no edges that have $b$ as an internal vertex, no new paths are obtained when we permit $b$ to be an internal vertex. Hence, $W_2 = W_1$.

$W_3$ has 1 as its $(i,j)$th entry if there is a path from $v_i$ to $v_j$ that has only $v_1 = a, v_2 = b$, and/or $v_3 = c$ as its internal vertices, if there is any. We now have paths from $d$ to $a$, namely, $dca$, and from $d$ to $d$, namely, $dcd$. Hence,

$$W_3 = \begin{pmatrix} 0 & 0 & 0 & 1 \\ 1 & 0 & 1 & 1 \\ 1 & 0 & 0 & 1 \\ 1 & 0 & 1 & 1 \end{pmatrix}.$$

Finally, $W_4$ has 1 as its $(i,j)$th entry if there is a path from $v_i$ to $v_j$ that has $v_1 = a$, $v_2 = b$, $v_3 = c$, and/or $v_4 = d$ as internal vertices, if there is any. Since they are all the vertices of the graph, this entry is 1 if and only if there is a path from $v_i$ to $v_j$. Hence,

$$W_4 = \begin{pmatrix} 1 & 0 & 1 & 1 \\ 1 & 0 & 1 & 1 \\ 1 & 0 & 1 & 1 \\ 1 & 0 & 1 & 1 \end{pmatrix}.$$

This last Matrix, $W_4$, is the matrix of the transitive closure.

Warshall's algorithm computes $M_{t(R)}$ by efficiently computing $W_0 = M_R, W_1, \cdots, W_n =$

$M_{t(R)}$. This observation shows that we can compute $W_k$ directly from $W_{k-1}$: There is a path from $v_i$ to $v_j$ with no vertices other than $v_1, v_2, \cdots, v_k$ as internal vertices if and only if either there is a path from $v_i$ to $v_j$ with its internal vertices among first $k-1$ vertices in the list, or there are paths from $v_i$ to $v_k$ and from $v_k$ to $v_j$ that have internal vertices only among the first $k-1$ vertices in the list. That is, either a path from $v_i$ to $v_j$ had already existed before $v_k$ was permitted as an internal vertex, or allowing $v_k$ as an internal vertex produces a path that goes from $v_i$ to $v_k$ and then from $v_k$ to $v_j$. The first type of path exists if and only if $w_{ij}^{(k-1)} = 1$, and the second type of path exists if and only if both $w_{ik}^{(k-1)}$ and $w_{kj}^{(k-1)}$ are 1. Hence, $w_{ij}^{(k)}$ is 1 if and only if either $w_{ij}^{(k-1)} = 1$ or both $w_{ik}^{(k-1)}$ and $w_{kj}^{(k-1)}$ are 1.

**LEMMA 4.4.2** Let $W_k = [w_{ij}^{(k)}]$ be the zero-one matrix that has a 1 in its $(i,j)$th position if and only if there is a path from $v_i$ to $v_j$ with internal vertices from the set $\{v_1, v_2, \cdots, v_k\}$. Then

$$w_{ij}^{(k)} = w_{ij}^{(k-1)} \vee (w_{ik}^{(k-1)} \wedge w_{kj}^{(k-1)})$$

whenever $i,j$ and $k$ are positive integers not exceeding $n$.

LEMMA 4.4.2 provides us a means to efficiently compute the matrices $W_k, k = 1, 2, \cdots, n$. We display the procedure as Algorithm 4.4.2.

**Algorithm 4.4.2 (Warshall's Algorithm)**
Procedure *Warshall* ($M_R$: $n \times n$ zero-one matrix)
    $W := M_R$
    for $k := 1$ to $n$
    Begin
        for $i := 1$ to $n$
        Begin
            for $j := 1$ to $n$
            $w_{ij} := w_{ij} \vee (w_{ik} \wedge w_{kj})$
        End
    End $\{W = [w_{ij}]$ is $M_{t(R)}\}$

It requires two bit operations using LEMMA 4.4.2 to find $w_{ij}^{(k)}$ from $w_{ij}^{(k-1)}$, $w_{ik}^{(k-1)}$, and $w_{kj}^{(k-1)}$, whereas finding all $n^2$ entries of $W_k$ from those of $W_{k-1}$ requires $2n^2$ bit operations. Since Warshall's algorithm begins with $W_0 = M_R$ and computes the sequences of $n$ zero-one matrices $W_1, W_2, \cdots, W_n = M_{t(R)}$, the total number of bit operations used is $n \cdot 2n^2 = 2n^3$.

During the computation of $W_k$, $w_{ij}^{(k)} = w_{ij}^{(k-1)} \vee (w_{ik}^{(k-1)} \wedge w_{kj}^{(k-1)})$ can be simplified to $w_{ij}^{(k)} = w_{ij}^{(k-1)} \vee w_{kj}^{(k-1)}$ when $w_{ik}^{(k-1)} = 1$, and $w_{ij}^{(k)} = w_{ij}^{(k-1)}$ otherwise.

**EXAMPLE 4.4.7** Suppose the relation matrix is $M_R = \begin{pmatrix} 1 & 0 & 0 & 1 \\ 0 & 1 & 1 & 1 \\ 0 & 0 & 1 & 0 \\ 1 & 1 & 0 & 0 \end{pmatrix}$, find $M_{t(R)}$.

*Solution*: $W_0 = M_R$. We now compute $W_1$ from $W_0$. Since $w_{11}^{(0)} = 1$ and $w_{41}^{(0)} = 1$, so for $j = 1,2,3,4$, $w_{1j}^{(1)} = w_{1j}^{(0)} \vee w_{1j}^{(0)} = w_{1j}^{(0)}$ and $w_{4j}^{(1)} = w_{4j}^{(0)} \vee w_{1j}^{(0)}$. Other entries do not change. Hence,

$$W_1 = \begin{pmatrix} 1 & 0 & 0 & 1 \\ 0 & 1 & 1 & 1 \\ 0 & 0 & 1 & 0 \\ 1 & 1 & 0 & 1 \end{pmatrix}.$$

Since $w_{22}^{(1)} = 1$, $w_{42}^{(1)} = 1$, we have that $w_{4j}^{(2)} = w_{4j}^{(1)} \vee w_{2j}^{(1)}$ for $j = 1,2,3,4$. And $w_{ij}^{(1)} = w_{ij}^{(2)}$ for $i = 1,2,3$; $j = 1,2,3,4$. Hence,

$$W_2 = \begin{pmatrix} 1 & 0 & 0 & 1 \\ 0 & 1 & 1 & 1 \\ 0 & 0 & 1 & 0 \\ 1 & 1 & 1 & 1 \end{pmatrix}.$$

Since $w_{23}^{(2)} = 1$, $w_{33}^{(2)} = 1$, and $w_{43}^{(2)} = 1$, we need to renew the values of the entries in the second and fourth row, and the results follow as $w_{2j}^{(3)} = w_{2j}^{(2)} \vee w_{3j}^{(2)}$ and $w_{4j}^{(3)} = w_{4j}^{(2)} \vee w_{3j}^{(2)}$ for $j = 1,2,3,4$, but $w_{1j}^{(3)} = w_{1j}^{(2)}$ and $w_{3j}^{(3)} = w_{3j}^{(2)}$ for $j = 1,2,3,4$. Hence,

$$W_3 = \begin{pmatrix} 1 & 0 & 0 & 1 \\ 0 & 1 & 1 & 1 \\ 0 & 0 & 1 & 0 \\ 1 & 1 & 1 & 1 \end{pmatrix}.$$

In the fourth column, $w_{14}^{(3)} = 1$, $w_{24}^{(3)} = 1$, and $w_{44}^{(3)} = 1$. We need to change the entries in the first and second row. For $j = 1,2,3,4$, $w_{1j}^{(4)} = w_{1j}^{(3)} \vee w_{4j}^{(3)}$, and $w_{2j}^{(4)} = w_{2j}^{(3)} \vee w_{4j}^{(3)}$. Thus,

$$W_4 = \begin{pmatrix} 1 & 1 & 1 & 1 \\ 1 & 1 & 1 & 1 \\ 0 & 0 & 1 & 0 \\ 1 & 1 & 1 & 1 \end{pmatrix}.$$

Therefore, $M_{t(R)} = W_4 = \begin{pmatrix} 1 & 1 & 1 & 1 \\ 1 & 1 & 1 & 1 \\ 0 & 0 & 1 & 0 \\ 1 & 1 & 1 & 1 \end{pmatrix}.$

Some properties are retained by closures. For example, we have the following results, the proof of which has been left as exercises to the readers:

### Inheritance Properties
1. If $R$ is reflexive, so are $s(R)$ and $t(R)$;
2. If $R$ is symmetric, so are $r(R)$ and $t(R)$;
3. If $R$ is transitive, so is $r(R)$.

## WORDS AND EXPRESSIONS

reflexive closure      自反闭包
symmetric closure      对称闭包
transitive closure      传递闭包

## EXERCISES 4.4

1. Find the symmetric closure of each of the following relations over the set $\{a, b, c, d\}$.
   a. $\{(a,b),(a,c),(b,c)\}$
   b. $\{(a,b),(b,a)\}$
   c. $\{(a,b),(b,c),(c,d),(d,a)\}$

2. Find the transitive closure of each of the relations in Exercise 1.

3. Draw the directed graphs of the reflexive closures of the relations with the directed graphs shown below:

4. Find the directed graphs of the symmetric closures of the relations with directed graphs shown in Exercises 3.

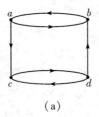 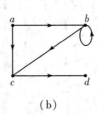 

(a)   (b)   (c)

5. Find the directed graphs of the transitive closures of the relations with directed graphs shown in Exercises 3.

6. Use Algorithm 4.4.1 to find the transitive closures of the following relations on $\{1, 2, 3, 4\}$.
   a. $\{(1, 2), (2, 1), (2, 3), (3, 4), (4, 1)\}$
   b. $\{(2, 1), (2, 3), (3, 1), (3, 4), (4, 1), (4, 3)\}$
   c. $\{(1, 2), (1, 3), (1, 4), (2, 3), (2, 4), (3, 4)\}$

7. Use Algorithm 4.4.2 to find the transitive closures of the following relations on $\{a,b,c,d,e\}$.
   a. $\{(a,c),(b,d),(c,a),(d,b),(e,d)\}$
   b. $\{(b,c),(b,e),(c,e),(d,a),(e,b),(e,c)\}$
   c. $\{(a,b),(a,c),(a,e),(b,a),(b,c),(c,a),(c,b),(d,a),(e,d)\}$

8. Prove LEMMA 4.4.1.

9. Prove each of the following statements about a relation $R$ over a set $A$.
   a. If $R$ is reflexive, then $s(R)$ and $t(R)$ are reflexive.
   b. If $R$ is symmetric, then $r(R)$ and $t(R)$ are symmetric.
   c. If $R$ is transitive, then $r(R)$ is transitive.

10. Prove each of the following statements about a binary relation $R$ over a set $A$.
    a. $rs(R) = sr(R)$
    b. $rt(R) = tr(R)$
    c. $st(R) \subseteq ts(R)$

## 4.5  Equivalence Relations

In this section we will study those relations that have a particular combination of properties that allow them to relate objects that are similar in some way.

**DEFINITION 4.5.1** A relation on a set $A$ is called an **equivalence relation** if it is reflexive, symmetric, and transitive.

Two elements that are related by an equivalence relation are called to be **equivalent**. This definition makes sense since an equivalence relation is symmetric. Also, since an equivalence relation is reflexive, in an equivalence relation every element is equivalent to itself. Furthermore, since an equivalence relation is transitive, if $a$ and $b$ are equivalent and $b$ and $c$ are equivalent, then $a$ and $c$ are equivalent.

**EXAMPLE 4.5.1** If $A = \{1,2,3\}$, then
$$R_1 = \{(1, 1), (2, 2), (3, 3)\}$$
$$R_2 = \{(1, 1), (2, 2), (2, 3), (3, 2), (3, 3)\}$$
$$R_3 = \{(1, 1), (1, 3), (2, 2), (3, 1), (3, 3)\}$$
$$R_4 = \{(1, 1), (1, 2), (1, 3), (2, 1), (2, 2), (2, 3), (3, 1), (3, 2), (3, 3)\} = A \times A$$
are all equivalence relations on $A$.

**EXAMPLE 4.5.2** Suppose that $R$ is the relation on the set of strings of English letters such that $aRb$ if and only if $l(a) = l(b)$, where $l(x)$ is the length of the string $x$. Is $R$ an equivalence relation?

*Solution*: Since $l(a) = l(a)$, it follows that $aRa$ whenever $a$ is a string, which means that $R$ is reflexive. Next, suppose that $aRb$. Then $l(a) = l(b)$. Therefore $bRa$, since $l(b) = l(a)$. Hence, $R$ is symmetric. Finally, suppose that $aRb$ and $bRc$. Then $l(a) = l(b)$ and $l(b) = l(c)$. Hence, $l(a) = l(c)$, and so $aRc$, which means that $R$ is transitive. Since $R$ is reflexive, symmetric, and transitive, it is an equivalence relation.

**EXAMPLE 4.5.3** Let $R$ be the relation on the set of real numbers such that $aRb$ if and only if $a - b$ is an integer. Is $R$ an equivalence relation?

*Solution*: Since $a - a = 0$ is an integer for any real number $a$, $aRa$ holds. Hence, $R$ is reflexive. Now suppose that $aRb$. Then $a - b$ is an integer, so that $b - a$ is also an integer. Hence $bRa$. It follows that $R$ is symmetric. If $aRb$ and $bRc$, then $a - b$ and $b - c$ are integers. Therefore, $a - c = (a - b) + (b - c)$ is also an integer. Hence $aRc$. Thus, $R$ is transitive. Consequently, $R$ is an equivalence relation.

**EXAMPLE 4.5.4** (Congruence Modulo m) Let $m$ be a positive integer and $m>1$. Show that the relation $R = \{(a,b) \mid a \equiv b(\bmod m)\}$ is an equivalence relation on the set of integers.

*Solution*: $a \equiv b(\bmod m)$ if and only if $m$ divides $a-b$. Note that $a-a=0$ is divisible by $m$, since $0 = 0 \cdot m$. Hence, $a \equiv a(\bmod m)$, so that congruence modulo $m$ is reflexive. Now suppose that $a \equiv b(\bmod m)$. Then $a-b$ is divisible by $m$, so that $a-b = km$, where $k$ is an integer. It follows that $b-a = (-k)m$, so that $b \equiv a(\bmod m)$. Hence, congruence modulo $m$ is symmetric. Finally, suppose that $a \equiv b(\bmod m)$ and $b \equiv c(\bmod m)$. Then $m$ divides both $a-b$ and $b-c$. Therefore, there are integers $k$ and $l$ such that $a-b = km$ and $b-c = lm$. Adding these two equations gives that $a-c = (k+l)m$. Thus, $a \equiv c(\bmod m)$. Therefore, congruence modulo m is transitive. It follows that congruence modulo m is an equivalence relation.

Let $A$ be the set of all students in your college. Consider the relation $R$ on $A$ that consists of all pairs $(a,b)$ where $a$ and $b$ have the same major. Given a student $a$, we can form the set of all students equivalent to $a$ with respect to $R$. This set consists of all students with the same major as $a$. This subset of $A$ is called an equivalence class of the relation.

**DEFINITION 4.5.2** Let $R$ be an equivalence relation on a set $A$. The set of all elements that are related to an element $a$ of $A$ is called the equivalence class of $a$. The **equivalence class** of $a$ with respect to $R$ is denoted by $[a]_R$. When only one relation is under consideration, we omit the subscript $R$ and write $[a]$ for this equivalence class.

In other words, if $R$ is an equivalence relation on a set $A$, the equivalence class of the element $a$ is $[a]_R = \{x \mid (a,x) \in R\}$. If $b \in [a]_R$, then $b$ is called a **representative** of the equivalence class. Any element of the class can be used as a representative, i. e., there is nothing special about the particular element chosen as the representative of the class.

**EXAMPLE 4.5.5** Find the equivalence class of an integer for the equivalence relation $R = \{(a,b) \mid a^2 = b^2\}$ on the set **Z**.

*Solution*: It is easy to find that $[0] = \{0\}$, $[1] = [-1] = \{1, -1\}$, $[2] = [-2] = \{2, -2\}$, and, in general, for any $n \in \mathbf{Z}$, $n \neq 0$, $[n] = [-n] = \{n, -n\}$.

**EXAMPLE 4.5.6** What are the equivalence classes for congruence modulo 4?

*Solution*: For relation $R = \{(a,b) \mid a \equiv b(\bmod 4)\}$, we find that
$[0] = \{a \mid a \equiv 0(\bmod 4)\} = \{\cdots, -8, -4, 0, 4, 8, 12, \cdots\} = \{4k \mid k \in \mathbf{Z}\}$;
$[1] = \{a \mid a \equiv 1(\bmod 4)\} = \{\cdots, -7, -3, 1, 5, 9, 13, \cdots\} = \{4k+1 \mid k \in \mathbf{Z}\}$;
$[2] = \{a \mid a \equiv 2(\bmod 4)\} = \{\cdots, -10, -6, -2, 2, 6, 10, 14, \cdots\} = \{4k+2 \mid k \in \mathbf{Z}\}$;
$[3] = \{a \mid a \equiv 3(\bmod 4)\} = \{\cdots, -5, -1, 3, 7, 11, 15, \cdots\} = \{4k+3 \mid k \in \mathbf{Z}\}$.

EXAMPLE 4.5.6 can be easily generalized by replacing 4 with any positive integer $m$. The equivalence classes of the relation congruence modulo $m$ are called the congruence classes module $m$. The congruence class of an integer $a$ modulo $m$ is denoted by $[a]_m$, i.e., $[a]_m = \{\cdots, a-2m, a-m, a, a+m, a+2m, \cdots\}$. It is easy to see that $\mathbf{Z} = [0] \cup [1] \cup [2] \cup [3]$. It means that the equivalence relation $R$ for congruence modulo 4 splits all integers into a collection of disjoint subsets, where each subset contains integers that are related to a specified integer. For instance, the subset $[1]$ contains all integers related to 1. This example illustrates how the equivalence classes of an equivalence relation partition a set into disjoint, nonempty subsets. We will make these notions more precise in the following discussion.

**THEOREM 4.5.1** Let $R$ be an equivalence relation on a set $A$, $a$ and $b$ are elements of $A$. The following statements are equivalent:
1. $aRb$
2. $[a] = [b]$
3. $[a] \cap [b] \neq \varnothing$

*Proof*: We first show that (1) implies (2). Assume that $aRb$. We prove $[a] = [b]$ by showing $[a] \subseteq [b]$ and $[b] \subseteq [a]$. Suppose $c \in [a]$. Then $aRc$. Since $aRb$ and $R$ is symmetric, we know that $bRa$. Furthermore, since $R$ is transitive and $bRa$ and $aRc$, it follows that $bRc$. Hence $c \in [b]$. This shows that $[a] \subseteq [b]$. Similarly, we can prove that $[b] \subseteq [a]$.

Secondly, we show that (2) implies (3). Assume that $[a] = [b]$. Since $R$ is reflexive, $a \in [a]$, i.e., $[a]$ is nonempty. It follows that $[a] \cap [b] = [a] \neq \varnothing$.

Finally, we show that (3) implies (1). Suppose that $[a] \cap [b] \neq \varnothing$. Then there is an element $c$ such that $c \in [a]$ and $c \in [b]$, i.e., $aRc$ and $bRc$. By the symmetric property, $cRb$. Since $aRc$ and $cRb$, we have $aRb$.

Since (1) implies (2), (2) implies (3), and (3) implies (1), the three statements

(1), (2) and (3) are equivalent.

Now we will show how an equivalence relation partition a set. Let $R$ be an equivalence relation over a set $A$. The union of the equivalence classes of $R$ is $A$. Since an element $a$ of $A$ is in its own equivalence class, namely, $[a]_R$, $\bigcup_{a \in A}[a]_R = A$.

In addition, it follows from THEOREM 4.5.1 that these equivalence classes are either equal or disjoint, so $[a]_R \cap [b]_R = \emptyset$ when $[a]_R \neq [b]_R$.

These observations show that the equivalence classes form a partition of $A$, since they split $A$ into disjoint subsets. More precisely, a **partition** of a set $S$ is a collection of disjoint nonempty subsets of $S$ that have $S$ as their union. In other words, the collection of subsets $A_i, i \in I$, (where $I$ is an index set) forms a partition of $S$ if and only if

$A_i \neq \emptyset, i \in I$

$A_i \cap A_j = \emptyset, i \neq j$

$\bigcup_{i \in I} A_i = S$.

**EXAMPLE 4.5.7** If $A = \{1, 2, 3, \cdots, 10\}$, then each of the following determines a partition of $A$:
1. $A_1 = \{1, 2, 3, 4, 5\}$, $A_2 = \{6, 7, 8, 9, 10\}$
2. $A_1 = \{1, 3, 5, 7, 9\}$, $A_2 = \{2, 4, 6, 8, 10\}$
3. $A_1 = \{1, 2, 3, 4\}$, $A_2 = \{5, 6, 7\}$, $A_3 = \{8, 9, 10\}$
4. $A_i = \{i, i+5\}, 1 \leq i \leq 5$.

**EXAMPLE 4.5.8** Let $A = \mathbf{R}$, and for each $i \in \mathbf{Z}$, let $A_i = [i, i+1)$. Then $\{A_i\}_{i \in \mathbf{Z}}$ is a partition of $R$.

**EXAMPLE 4.5.9** If $A = \{1, 2, 3, 4, 5\}$ and $R = \{(1, 1), (2, 2), (2, 3), (3, 2), (3, 3), (4, 4), (4, 5), (5, 4), (5, 5)\}$ then $R$ is an equivalence relation on $A$. Obviously $[1] = \{1\}$, $[2] = \{2, 3\} = [3]$, $[4] = \{4, 5\} = [5]$, and $A = [1] \cup [2] \cup [4]$ with $[1] \cap [2] = \emptyset$, $[2] \cap [4] = \emptyset$, and $[1] \cap [4] = \emptyset$. So $\{[1], [2], [4]\}$ determines a partition of $A$.

We have seen that the equivalence classes of an equivalence relation on a set form a partition of the set. The subsets in this partition are the equivalence classes. Conversely, every partition of a set can be used to form an equivalence relation. Two elements are equivalent with respect to this relation if and only if they are in the same subset of the partition. To see this, assume that $\{A_i | i \in I\}$ is a partition of a set $S$. Let $R$ be the relation on $S$ consisting of the pairs $(x,y)$ where $x$ and $y$ belong to the same subset $A_i$ in the partition. To show that $R$ is an equivalence relation we must show that $R$ is reflexive, symmetric, and transitive.

First, $(a,a) \in R$ for all $a \in S$, since $a$ is in the same subset of itself. Hence, $R$ is reflexive. Next, if $(a,b) \in R$, then $b$ and $a$ are in the same subset of the partition, so that $(b,a) \in R$ as well, which means that $R$ is symmetric. Finally, if $(a,b) \in R$ and $(b,c) \in R$, then $a$ and $b$ are in the same subset $A$, and $b$ and $c$ are in the same subset $B$. Since the subsets of the partition are disjoint, and $b$ belongs to $A$ and $B$, it follows that $A = B$. Consequently, $a$ and $c$ belong to the same subset of the partition, which leads to $(a,c) \in R$. Thus, $R$ is transitive. It follows that $R$ is an equivalence relation. The equivalence classes of $R$ are the subsets of $S$ containing related elements, and by the definition of $R$, these are the subsets of the partition. This gives us the following theorem.

**THEOREM 4.5.2** If $S$ is a set, then
1. for any equivalence relation $R$ on $S$, the equivalence classes of $R$ form a partition of $S$; and conversely,
2. for any partition $\{A_i | i \in I\}$ of $S$, there is an equivalence relation $R$ that has the sets $A_i, i \in I$, as its equivalence classes.

**EXAMPLE 4.5.10** If an equivalence relation $R$ on $A = \{1, 2, 3, 4, 5, 6, 7\}$ induces the partition $A = \{1, 2\} \cup \{3\} \cup \{4, 5, 7\} \cup \{6\}$, what is $R$ ?
*Solution:* Consider the subset $\{1, 2\}$ of the partition. This subset implies that $[1] = [2] = \{1, 2\}$, and so $(1, 1), (2, 2), (1, 2), (2, 1) \in R$. Similarly, the subset $\{4, 5, 7\}$ implies that under $R$, $[4] = [5] = [7] = \{4,5,7\}$. As an equivalence relation, $R$ must contain $\{4,5,7\} \times \{4,5,7\} = \{(4,4), (4,5), (4,7), (5,4), (5,5), (5,7), (7,4), (7,5), (7,7)\}$. In fact, $R = \{\{1,2\} \times \{1,2\}\} \cup \{\{3\} \times \{3\}\} \cup \{\{4,5,7\} \times \{4,5,7\}\} \cup \{\{6\} \times \{6\}\}$.

**EXAMPLE 4.5.11** List the ordered pairs in the equivalence relation $R$ on $S = \{1, 2, 3, 4, 5\}$ produced by the partition $A_1 = \{1, 2\}$, $A_2 = \{3, 4\}$, $A_3 = \{5\}$.
*Solution:* Since $A_1 \times A_1 = \{(1, 1), (1, 2), (2, 1), (2, 2)\}$, $A_2 \times A_2 = \{(3, 3), (3, 4), (4, 3), (4, 4)\}$, and $A_3 \times A_3 = \{(5, 5)\}$, we obtain that $R = \{(1, 1), (1, 2), (2, 1), (2, 2), (3, 3), (4, 4), (3, 4), (4, 3), (5, 5)\}$.

**EXAMPLE 4.5.12** Suppose that $A = \{1, 2, 3, 4, 5, 6, 7\}$, $B = \{x, y, z\}$, $f: A \to B$ is the onto function defined by $f = \{(1, x), (2, z), (3, x), (4, y), (5, z), (6, y), (7, x)\}$. The relation $R$ defined on $A$ is that $aRb$ if $f(a) = f(b)$. It can be shown that $R$ is an equivalence relation. Here

$f^{-1}(x) = \{1, 3, 7\} = [1] = [7],$
$f^{-1}(y) = \{4, 6\} = [4] = [6],$
$f^{-1}(z) = \{2, 5\} = [2] = [5].$

Since $A = [1] \cup [4] \cup [2] = f^{-1}(x) \cup f^{-1}(y) \cup f^{-1}(z)$, we see that $\{f^{-1}(x), f^{-1}(y), f^{-1}(z)\}$ determines a partition of $A$.

In fact, for any nonempty sets $A$ and $B$, if $f: A \to B$ is an onto function, then $A = \bigcup_{b \in B} f^{-1}(b)$. Therefore $\{f^{-1}(b) \mid b \in B\}$ forms a partition of $A$.

## WORDS AND EXPRESSIONS

| | |
|---|---|
| equivalence relation | 等价关系 |
| congruence modulo $m$ | 模 $m$ 同余 |
| equivalence class | 等价类 |
| representative | 代表 |
| partition | 划分 |

## EXERCISES 4.5

1. Which of these relations on $\{0, 1, 2, 3\}$ are equivalence relations? Determine the properties of an equivalence relation that the others lack.
    a. $\{(0, 0), (1, 1), (2, 2), (3, 3)\}$
    b. $\{(0, 0), (0, 2), (2, 0), (2, 2), (2, 3), (3, 2), (3, 3)\}$
    c. $\{(0, 0), (1, 1), (1, 2), (2, 1), (2, 2), (3, 3)\}$
    d. $\{(0, 0), (1, 1), (1, 3), (2, 2), (2, 3), (3, 1), (3, 2), (3, 3)\}$
    e. $\{(0, 0), (0, 1), (0, 2), (1, 0), (1, 1), (1, 2), (2, 0), (2, 2), (3, 3)\}$

2. For each of the following relations, either prove it is an equivalence relation or prove it is not.
    a. $R_1 = \{(a, b) \mid a + b \text{ is even}\}$ on the set of integers.
    b. $R_2 = \{(a, b) \mid ab > 0\}$ on the set of nonzero rational numbers.

c. $R_3 = \{(a,b) \mid a - b \text{ is an integer}\}$ on the set of rational numbers.
d. $R_4 = \{(a,b) \mid |a - b| \leq 2\}$ on the set of natural numbers.

3. Determine whether the relations with directed graphs shown below are equivalence relations.

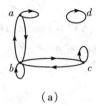
(a)
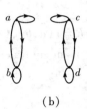
(b)
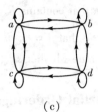
(c)

4. Determine whether the relations represented by these zero-one matrices are equivalence relations.

a. $\begin{pmatrix} 1 & 1 & 1 \\ 0 & 1 & 1 \\ 1 & 1 & 1 \end{pmatrix}$
b. $\begin{pmatrix} 1 & 0 & 1 & 0 \\ 0 & 1 & 0 & 1 \\ 1 & 0 & 1 & 0 \\ 0 & 1 & 0 & 1 \end{pmatrix}$
c. $\begin{pmatrix} 1 & 1 & 1 & 0 \\ 1 & 1 & 1 & 0 \\ 1 & 1 & 1 & 0 \\ 0 & 0 & 0 & 1 \end{pmatrix}$

5. Determine whether each of the following collections of sets is a partition of $\{a,b,c,d,e,f,g\}$.
    a. $\{a,b,c\}, \{d,f\}, \{e,g\}$
    b. $\{d,e\}, \{a,c,d\}, \{g\}, \{h\}$
    c. $\{a,b,c\}, \{b,c,d,e\}, \{f,g\}$
    d. $\{a,c,e\}, \{d,f,g\}$

6. Which of these are partitions of the set of real numbers?
    a. The set of intervals $[k, k+1], k = \cdots, -2, -1, 0, 1, 2, \cdots$
    b. The set of intervals $(k, k+1), k = \cdots, -2, -1, 0, 1, 2, \cdots$
    c. The set of intervals $(k, k+1], k = \cdots, -2, -1, 0, 1, 2, \cdots$
    d. The sets $\{x + n \mid n \in \mathbf{Z}\}$, for all $x \in [0, 1)$

7. List the ordered pairs in the equivalence relations produced by the following partitions of $\{a,b,c,d,e,f\}$.
    a. $\{a,b\}, \{c,d\}, \{e,f\}$
    b. $\{a\}, \{b\}, \{c,d\}, \{e,f\}$

c. $\{a,c,e,f\}, \{b,d\}$

8. Find the smallest equivalence relation on the set $\{a, b, c, d, e\}$ containing the relation $\{(a,b),(a,c),(d,e)\}$.

9. Show that if $R$ is a relation over $A$, then $t(s(r(R)))$ is the smallest equivalence relation that contains $R$.

10. Suppose that $A = \{a,b,c\}$ and $R = \{(a,b),(a,c),(b,b)\}$. Find $s(t(r(R)))$. Determine whether $s(t(r(R)))$ is an equivalence relation on $A$.

## 4.6 Partial Orderings

We often use relations to order some or all of the elements of sets. For instance, in the construction of a house, certain jobs such as digging the foundation, must be performed before other phases of the construction can be undertaken. If $A$ is a set of tasks that must be performed in building a house, we can define a relation $R$ on $A$ by $xRy$ if $x,y$ denote the same task or if task $x$ must be performed before the start of task $y$. In this way we place an order on the elements of $A$. This is sometimes referred to as a PERT (Program Evaluation and Review Technique) network. (Such networks came into play during the 1950s in order to handle the complexities that arose in organizing many individual activities required for the completion of projects on a very large scale. This technique was actually developed and first used by the U. S. Navy in order to coordinate many projects that were necessary for building the Polaris submarine.) Note that $R$ is reflexive, antisymmetric and transitive. These are the properties that characterize the relations used to order the elements of sets.

Now let's discuss the basic ideas and techniques of ordering.

**DEFINITION 4.6.1** A relation $R$ on a set $S$ is called a **partial ordering** or **partial order** if it is reflexive, antisymmetric, and transitive. A set $S$ together with a partial ordering $R$ is called a **partially ordered set** or **poset**, and is denoted by $(S,R)$.

**EXAMPLE 4.6.1** The "greater than or equal to" relation ($\geq$) is a partial ordering on the set of integers. It follows that $(\mathbf{Z}, \geq)$ is a poset.

**EXAMPLE 4.6.2** Define $R$ on $A = \{1, 2, 3, 4\}$ by $xRy$ if $x|y$, i. e., $x$ (exactly) divides $y$. Then $R = \{(1, 1), (1, 2), (1, 3), (1, 4), (2, 2), (2, 4), (3, 3),$

$(4,4)\}$ is a partial order, and $(A,R)$ is a poset.

**EXAMPLE 4.6.3** Show that the inclusion relation $\subseteq$ is a partial order on the power set of a set $S$.

*Solution*: Since $A \subseteq A$ whenever $A$ is a subset of $S$, $\subseteq$ is reflexive. It is antisymmetric since $A \subseteq B$ and $B \subseteq A$ imply that $A = B$. Finally, $\subseteq$ is transitive, since $A \subseteq B$ and $B \subseteq C$ imply that $A \subseteq C$. Hence, $\subseteq$ is a partial order on $P(S)$, and $(P(S), \subseteq)$ is a poset.

**EXAMPLE 4.6.4** Consider the diagrams given in Figure 4.6.1. If diagram (a) was part of the directed graph associated with a relation $R$, then for $(1,2)$, $(2,1) \in R$ and $1 \neq 2$, $R$ could not be antisymmetric.

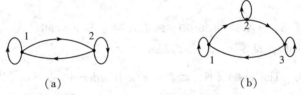

**Figure 4.6.1  Diagrams for EXAMPLE 4.6.4**

For part (b), if the diagram was part of the graph of a transitive relation $R$, then $(1,2)$, $(2,3) \in R$. Since $(3,1) \in R$, and $1 \neq 3$, $R$ could not be antisymmetric. So it is not a partial order.

From the observations in EXAMPLE 4.6.4, given a relation $R$ on a set $S$, assuming $G$ is the diagram associated with $R$, we find that

1. if $G$ contains a pair of edges of the form $(a,b)$ and $(b,a)$ where $a,b \in S$ but $a \neq b$, or
2. if $R$ is transitive and $G$ contains a directed cycle (of length greater than or equal to three), then the relation $R$ cannot be antisymmetric, so $(S,R)$ fails to be a poset.

When we talk about partial orders, we often use the notion $a \leqslant b$ to denote $(a,b) \in R$. We can read $a \leqslant b$ as $a$ is less than or equal to $b$. Note that the symbol $\leqslant$ is used to denote partial orders in any poset, not just the "less than or equal to" relation in normal sense. The notation $a < b$ denotes that $a \leqslant b$, but $a \neq b$. Also, we say "$a$ is less than $b$" or "$b$ is greater than $a$", if $a < b$.

When $a$ and $b$ are elements of a poset $(S, \leqslant)$, it is not necessary that either $a \leqslant b$ or $b \leqslant a$. For instance, in $(P(\mathbf{Z}), \subseteq)$, $\{1, 2\}$ is not related to $\{2, 3\}$, and vice versa. Similarly, in $(\mathbf{Z}, |)$, 2 is not related to 5 and nor is 5 related to 2. This leads to the following definition.

**DEFINITION 4.6.2** Two elements $a$ and $b$ of a poset $(S, \leqslant)$ are called **comparable** if either $a \leqslant b$ or $b \leqslant a$. When $a$ and $b$ are elements of $S$ such that neither $a \leqslant b$ nor $b \leqslant a$, $a$ and $b$ are called **incomparable**. If every two elements of $S$ are comparable, $\leqslant$ is called a **total order** or a **linear order**, and $(S, \leqslant)$ is called a **totally ordered** or **linearly ordered set**. A totally ordered set is also called a **chain**.

**EXAMPLE 4.6.5** The poset $(\mathbf{Z}, \leqslant)$ is totally ordered, but $(\mathbf{Z}^+, |)$ is not totally ordered.

**DEFINITION 4.6.3** A **well-ordered set** $(S, \leqslant)$ is a totally ordered set such that every nonempty subset of $S$ has a least element.

**EXAMPLE 4.6.6** The poset $(\mathbf{N}, \leqslant)$ is a well-ordered set, but $(\mathbf{Z}, \leqslant)$ is not a well-ordered set since the set of negative integers, which is a subset of $\mathbf{Z}$, has no least element.

Mathematical induction is usually used to prove the propositions of the form $\forall n\, P(n)$, where the universe of discourse is the set of positive integers. The principle of well-ordered induction is used to prove the propositions of the form $\forall x\, P(x), x \in S$ where $S$ is a well-ordered set.

The validity of mathematical induction follows from the following fundamental axiom about the set of integers.

**THE WELL-ORDERING PROPERTY** *Every nonempty set of nonnegative integers has a least element.*

Why is mathematical induction a valid proof technique? The reason comes from the well-ordering property. Suppose we know that $P(1)$ is true and the proposition $P(k) \rightarrow P(k+1)$ is true for all positive integers $k$. To show that $P(n)$ must be true for all positive integers $n$, assume that there is at least one positive integer $n$ for which $P(n)$ is false. Then the set $S$ of positive integers for which $P(n)$ is false is nonempty. Thus, by the well-ordering property, $S$ has a least element, which is denoted by $m$. We know that $m$ cannot be 1, because $P(1)$ is true. Since $m$ is positive and greater than 1, $m-1$ is a positive integer. Furthermore, since $m-1$ is less than $m$, it is not in $S$, so $P(m-1)$ must be true. Since the implication $P(m-1) \rightarrow P(m)$ is also true, it must be the case that $P(m)$ is true. This contradicts the choice of $m$. Hence, $P(n)$ must be true for every positive integer $n$.

We now state and prove the principle of well-ordered induction.

**THEOREM 4.6.1** (*The principle of well-ordered induction*) Suppose that $S$ is a well-ordered set. Then $P(x)$ is true for all $x \in S$, if
BASIS STEP: $P(x_0)$ is true for the least element $x_0$ of $S$, and
INDUCTION STEP: For every $y \in S$ if $P(x)$ is true for all $x < y$, then $P(y)$ is true.
*Proof*: Suppose the theorem is not true. Then there is an element $y \in S$ such that $P(y)$ is false. Consequently, the set $A = \{x \in S | P(x) \text{ is false}\}$ is nonempty. Since $S$ is a well-ordered set, the set $A$ has a least element $a$. We know that $a \neq x_0$, since by the basis step $P(x_0)$ is true. Because $a$ is the least element of $A$, we know that $P(x)$ is true for all $x \in S$ if $x < a$. This implies by the induction step that $P(a)$ is true. This contradiction shows that $P(x)$ must be true for all $x \in S$.

The principle of well-ordered induction is a versatile technique for proving propositions about well-ordered sets. Even when it is possible to use mathematical induction for the set of positive integers to prove a theorem, it may be simpler to use the principle of well-ordered induction.

As an example, we define an ordering on $\mathbf{N} \times \mathbf{N}$, the ordered pairs of nonnegative integers, by specifying that $(x_1, y_1)$ is less than or equal to $(x_2, y_2)$ if either $x_1 < x_2$, or $x_1 = x_2$ and $y_1 < y_2$; this is called the lexicographic ordering. $(\mathbf{N} \times \mathbf{N}, \leq)$ is a well-ordered set.

**EXAMPLE 4.6.7** Suppose that $a_{m,n}$ is defined recursively for $(m,n) \in \mathbf{N} \times \mathbf{N}$ by $a_{0,0} = 0$ and

$$a_{m,n} = \begin{cases} a_{m-1,n} + 1, & \text{if } n = 0 \text{ and } m > 0 \\ a_{m,n-1} + n, & \text{if } n > 0 \end{cases}.$$

Show that $a_{m,n} = m + \dfrac{n(n+1)}{2}$ for all $(m,n) \in \mathbf{N} \times \mathbf{N}$.

*Solution*: We can prove that $a_{m,n} = m + \dfrac{n(n+1)}{2}$ using the principle of the well-ordered induction. The basis step requires that we show that this formula is valid when $(m,n) = (0,0)$. The induction step requires that we show that if this formula holds for all pairs smaller than $(m,n)$ in the lexicographic ordering of $\mathbf{N} \times \mathbf{N}$, then it also holds for $(m,n)$.

*BASIS STEP*: Let $(m,n) = (0,0)$. Then by the basis case of the recursive definition of $a_{m,n}$, we have $a_{0,0} = 0$. Furthermore, when $m = n = 0$, $m + \frac{n(n+1)}{2} = 0$. This completes the basis step.

*INDUCTIVE STEP*: Suppose that $a_{m',n'} = m' + \frac{n'(n'+1)}{2}$ whenever $(m', n')$ is less than $(m,n)$ in the lexicographic ordering of $\mathbf{N} \times \mathbf{N}$.

By the recursive definition, if $n = 0$, then $a_{m,n} = a_{m-1,n} + 1$. Because $(m - 1, n)$ is smaller than $(m, n)$, the induction hypothesis tells us that $a_{m-1,n} = m - 1 + \frac{n(n+1)}{2}$, so that $a_{m,n} = m - 1 + \frac{n(n+1)}{2} + 1 = m + \frac{n(n+1)}{2}$, giving us the desired equality. Now suppose that $n > 0$. Then $a_{m,n} = a_{m,n-1} + n$. Since $(m, n - 1)$ is smaller than $(m, n)$, the inductive hypothesis tells us that $a_{m,n-1} = m + \frac{(n-1)n}{2}$. So $a_{m,n} = m + \frac{(n-1)n}{2} + n = m + \frac{n(n+1)}{2}$. This finishes the inductive step.

We will now discuss the **lexicographic order**. The words in a dictionary are listed in alphabetic or lexicographic order, which is based on the ordering of the letters in the alphabet. This is a special case of ordering of strings constructed from a partial ordering on the set. We will show how this construction works in any poset.

First, we will show how to construct a partial ordering on the Cartesian product of two posets, $(A_1, \preccurlyeq_1)$ and $(A_2, \preccurlyeq_2)$. The lexicographic ordering $\preccurlyeq$ on $A_1 \times A_2$ is defined as follows: $(a_1, a_2) \prec (b_1, b_2)$ if and only if either $a_1 \prec_1 b_1$ or both $a_1 = b_1$ and $a_2 \prec_2 b_2$. We then obtain a partial order $\preccurlyeq$ by adding equality to $\prec$ on $A_1 \times A_2$.

**EXAMPLE 4.6.8** Determine whether $(3, 6) \prec (5, 8)$, and whether $(5, 8) \prec (5, 11)$ in the poset $(\mathbf{Z} \times \mathbf{Z}, \preccurlyeq)$, where $\preccurlyeq$ is the lexicographic ordering constructed from the usual $\leq$ relation on $\mathbf{Z}$.

*Solution*: Since $3 < 5$, it follows that $(3, 6) < (5, 8)$. Also, we have $(5, 8) < (5, 11)$, since the first entries in $(5, 8)$ and $(5, 11)$ are the same, but $8 < 11$.

A lexicographic ordering can be defined on the Cartesian Product of $n$ posets
$$(A_1, \leqslant_1), (A_2, \leqslant_2), \cdots, (A_n, \leqslant_n).$$
Define the partial ordering $\leqslant$ on $A_1 \times A_2 \times \cdots \times A_n$ by $(a_1, a_2, \cdots, a_n) < (b_1, b_2, \cdots, b_n)$ if $a_1 <_1 b_1$, or if there is an integer $i > 0$ such that $a_1 = b_1, \cdots, a_i = b_i$, and $a_{i+1} <_{i+1} b_{i+1}$.

For example $(1, 2, 3, 4, 5) < (1, 2, 4, 3, 5)$, since the entries in the first two positions of these 5-tuples agree, but in the third position the entry in the first 5-tuple, 3, is less than that in the second 5-tuple, 4.

We can now define lexicographic ordering of strings. Consider the strings $a_1, a_2, \cdots, a_m$ and $b_1, b_2, \cdots, b_n$ on a partially ordered set $S$. Suppose these strings are not equal. Let $t = \min\{m, n\}$. The definition of lexicographic ordering is that the string $a_1, a_2, \cdots, a_m$ is less than $b_1, b_2, \cdots, b_n$ if and only if
$(a_1, a_2, \cdots, a_t) < (b_1, b_2, \cdots, b_t)$, or
$(a_1, a_2, \cdots, a_t) = (b_1, b_2, \cdots, b_t)$ and $m < n$.

**EXAMPLE 4.6.9** Consider the set of strings of lower case English letters. Using the ordering of letters in the alphabet, a lexicographic ordering on the set of strings can be constructed. A string is less than a second string if the letter in the first string in the first position where the strings differ comes before the letter in the second string in this position, or if the first string and the second string agree in all positions, but the second string has more letters. This ordering is the same as that used in dictionaries. For example, *abcdef* < *abecdf*, since these strings differ first in the third position, and $c < e$. Also, *abcdef* < *abcdefg*, since the first six letters agree, but the second string is longer.

**EXAMPLE 4.6.10** Consider the directed graph for the partial order $R = \{(1, 1), (2, 2), (3, 3), (4, 4), (1, 2), (2, 4), (1, 4), (1, 3)\}$ on set $\{1, 2, 3, 4\}$. Figure 4.6.2(a) is the graphical representation of $R$. In part (b) of this figure, we have a somewhat simpler diagram, which is called the Hasse diagram for $R$.

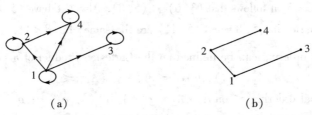

(a)            (b)

**Figure 4.6.2**    The directed graph and the Hasse diagram for EXAMPLE 4.6.10

When we know that a relation $R$ is a partial ordering on a set, we can eliminate the loops at the vertices of its directed graph. Since $R$ is also transitive, having the edges $(1, 2)$ and $(2, 4)$ is enough to insure the existence of edge $(1, 4)$, so we don't need to include that edge. In this way we obtain the diagram in Figure 4.6.2(b).

In general, we can represent a partial ordering on a finite set using this procedure: Start with the directed graph for this relation. Because a partial order is reflexive, a loop is present at every vertex. Remove these loops. Remove all edges that must be present because of the transitivity, since they should naturally be there because a partial ordering is transitive. For instance, If $(a,b)$ and $(b,c)$ are in the partial ordering, remove the edge $(a,c)$, since it must be present. Furthermore, if $(c,d)$ is also in the partial order, remove the edge $(a,d)$, since it must be present as well. Finally, arrange each edge so that its initial vertex is below its terminal point, i.e., all edges are "upward" toward their terminal vertices.

These steps are well defined, and only a finite number of steps need to be carried out for a finite poset. When all the steps have been taken, the resulting diagram contains sufficient information to find the partial ordering. This diagram is called a **Hasse diagram**.

**EXAMPLE 4.6.11**    Draw the Hasse diagrams for the following four posets.
1. $(P(A), \subseteq)$, $A = \{a,b,c\}$
2. $(A, |)$, $A = \{3,6,12,24\}$
3. $(A, |)$, $A = \{2,3,5,7\}$
4. $(A, |)$, $A = \{2,3,5,6,7,11,12,35,385\}$

*Solution*: The Hasse diagrams are shown in Figure 4.6.3.

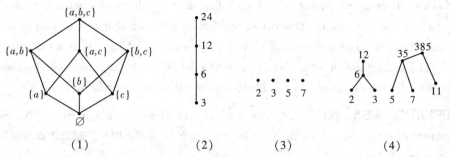

**Figure 4.6.3** The Hasse diagrams for EXAMPLE 4.6.11

When we have a partially ordered set, it's natural to use words like "minimal", "least", "maximal", and "greatest". Elements that have these extremal properties are important for many applications. Let's now give these words some formal definitions.

**DEFINITION 4.6.4** If $(S, \leqslant)$ is a poset, then an element $a$ is called a **maximal** element if there is no $b \in S$ such that $a < b$. An element $a$ is called a **minimal** element if there is no $b \in S$ such that $b < a$.

**EXAMPLE 4.6.12** Which elements of the poset $(\{2, 3, 4, 5, 10, 12, 35\}, |)$ are maximal, and which are minimal?

*Solution*: The Hasse diagram for this poset is shown in Figure 4.6.4, which shows that 10, 12, and 25 are maximal elements, and the minimal elements are 2, 3 and 5.

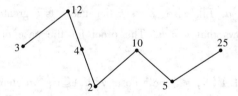

**Figure 4.6.4** The Hasse diagram for EXAMPLE 4.6.12

As this example shows, a poset can have more than one maximal element and more than one minimal element. Are there any conditions indicating when a poset must have a maximal or minimal element? Let's answer this question by the following theorem.

**THEOREM 4.6.2** If $(S, \leqslant)$ is a poset and $S$ is finite, then $S$ has both a maximal and a minimal element.

*Proof*: Let $a_1 \in S$. If there is no element $a \in S$ such that $a_1 < a$, then $a_1$ is maximal.

Otherwise there is an element $a_2 \in S$ such that $a_1 < a_2$. If no element $a \in S$ satisfies $a_2 < a$, then $a_2$ is maximal. Otherwise, we find $a_3 \in S$ such that $a_1 < a_2$ and $a_2 < a_3$. Continuing in this manner, since $S$ is finite, we get to an element $a_n \in S$ such that no element $a \in S$ satisfies $a_n < a$. According to the definition, $a_n$ is a maximal element. The proof for the existence of a minimal element follows in a similar way.

**DEFINITION 4.6.5** Let $(S, \leqslant)$ be a poset. An element $a \in S$ is called **the least element** if $a \leqslant b$ for all $b \in S$. An element $a \in S$ is called **the greatest element** if $b \leqslant a$ for all $b \in S$.

**EXAMPLE 4.6.13** For the partial orders in EXAMPLE 4.6.11, we find that

1. The partial order in part (1) has a least element $\emptyset$ and a greatest element $\{a, b, c\}$;
2. The partial order in part (2) has a least element 3 and a greatest element 24;
3. There is no greatest element or least element for the poset in part (3);
4. No greatest element nor least element exists for the partial order in part (4).

We have seen that it is possible for a poset to have several maximal and minimal elements. What about least and greatest elements?

**THEOREM 4.6.3** If the poset $(S, \leqslant)$ has a greatest (least) element, then that element is unique.

*Proof*: Suppose that $a, b \in S$ and that both $a$ and $b$ are greatest elements. Since $a$ is a greatest element, $b \leqslant a$. Likewise, $a \leqslant b$ because $b$ is a greatest element. As $\leqslant$ is antisymmetric, it follows that $a = b$. The proof for the case of the least element is similar.

**DEFINITION 4.6.6** Let $(S, \leqslant)$ be a poset with $A \subseteq S$. An element $u \in S$ is called an **upper bound** of $A$ if $a \leqslant u$ for all $a \in A$. Likewise, an element $l \in S$ is called a **lower bound** of $A$ if $l \leqslant a$ for all $a \in A$.

An element $x \in S$ is called the **least upper bound** of $A$ if $x$ is a upper bound of $A$ and for all of other upper bound $u$ of $A$, $x \leqslant u$. Similarly, $y \in S$ is called the **greatest lower bound** of $A$ if $y$ is a lower bound of $A$ and for all of other lower bound $l$ of $A$, $l \leqslant y$.

**EXAMPLE 4.6.14** Find the lower and upper bounds of the subsets $\{1, 2, 3\}$, $\{6, 8\}$, and $\{1, 3, 4, 6\}$ in the poset with the Hasse diagram shown in Figure 4.6.5. Point out the greatest lower bound and the least upper bound of $\{2, 5, 8\}$ if they exist.

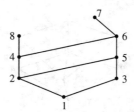

**Figure 4.6.5** The Hasse diagram for EXAMPLE 4.6.14

*Solution*: The upper bounds of $\{1, 2, 3\}$ are 5, 6, 7, and its only lower bound is 1. There are no upper bounds of $\{6, 8\}$, and its lower bounds are 4, 2, and 1. The upper bounds of $\{1, 3, 4, 6\}$ are 6 and 7, and its lower bound is 1. There are no upper bounds of $\{2, 5, 8\}$. Hence there does not exist the least upper bound of $\{2, 5, 8\}$. Its lower bounds are 2 and 1. Therefore, 2 is the greatest lower bound.

**DEFINITION 4.6.7** A poset $(S, \leq)$ is called a **lattice** if for any $a, b \in S$, the subset $\{a, b\}$ has both the least upper bound and the greatest lower bound.

**EXAMPLE 4.6.15** For $S = \mathbf{Z}^+$ and $a, b \in \mathbf{Z}^+$, define $(a, b) \in R$ if and only if $a \mid b$, then $(\mathbf{Z}^+, \mid)$ is a lattice. The least upper bound for positive integers $a$ and $b$ is the least common multiple and the greatest lower bound is the greatest common divisor of $a$ and $b$.

**EXAMPLE 4.6.16** For a set $S$, $(P(S), \subseteq)$ is a lattice. Since the greatest lower bound for two subsets $A$ and $B$ of $S$ is $A \cap B$, and the least upper bound of $A$ and $B$ is $A \cup B$.

**EXAMPLE 4.6.17** Determine whether the poset represented by the Hasse diagram in Figure 4.6.6 is a lattice.

*Solution*: The poset represented by the Hasse diagram shown in Figure 4.6.6 is not a lattice, since $\{2, 3\}$ does not have the least upper bound.

Suppose that a manufacturer is about to market a new product. The new product is made by finishing 7 different tasks, denoted by $a, b, c, d, e, f$ and $g$, some of which can be started only after others have been finished. How can an order be found for these tasks? To model this problem we set up a partial order on the set of tasks, so that $x < y$ if and only if $x$ and $y$ are tasks where $y$ cannot be started until $x$ has been completed. Figure 4.6.7 shows the partial order for the new product.

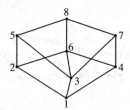

**Figure 4.6.6** The Hasse diagram for EXAMPLE 4.6.17

**Figure 4.6.7** The Hasse diagram for the new product

To produce a schedule for the manufacturer, we need to order the 7 tasks and make sure that the partial order of the Hasse diagram is maintained. What we are really asked for here is that we should find a total order $S$ on these tasks such that the original partial order given by the Hasse diagram is a subset of $S$.

The technique that we need to do this job is **topological sorting**. To define a total ordering on a given poset $(A, R)$ with $|A| = n$, first choose a minimal element $a_1$; such an element exists according to THEOREM 4.6.2. Next, note that $(A - \{a_1\}, R)$ is also a poset (the proof is left to the readers). If it is nonempty, choose a minimal element $a_2$ of this poset. Then choose a minimal element $a_3$ in $A - \{a_1, a_2\}$, if $A - \{a_1, a_2\}$ is nonempty. Continue this process by choosing $a_{k+1}$ to be a minimal element in $A - \{a_1, a_2, \cdots, a_k\}$ as long as there are elements remain. If $a_n$ has been chosen, the process is completed and we have a total order $S$:

$$a_1 \leqslant a_2 \leqslant a_3 \leqslant \cdots \leqslant a_n$$

which contains $R$.

The Pseudo code for this topological sorting algorithm is shown in Algorithm 4.6.1.

**Algorithm 4.6.1** Topological Algorithm

    Procedure *topological sort* ($A$: finite set)

    $k := 1$

while $A \neq \emptyset$

begin

$\quad a_k :=$ a minimal element of $A$

$\quad A := A - \{a_k\}$

$\quad k := k+1$

End $\{a_1, a_2, a_3, \cdots, a_n$ is a total order of $A\}$

**EXAMPLE 4.6.18** Find a total order for the poset shown in Figure 4.6.7

*Solution*: The first step is to find a minimal element. There are two minimal elements $a$ and $b$, and we choose $a$. Next, find a minimal element in $\{b,c,d,e,f,g\}$, and this must be $b$, since it is the only minimal element. $c$ is the minimal element of $\{c,d,e,f,g\}$. At this stage, $d$ and $g$ are minimal elements. We select $d$. Since both $e$ and $g$ are minimal elements of $\{e,f,g\}$, either of them can be chosen next. We select $g$. The remaining elements are $e$ and $f$. Both $e$ and $f$ are minimal elements, and we choose $e$, which leaves $f$ as the last element left. This produces the total order $a < b < c < d < g < e < f$.

The steps used by this sorting algorithm are displayed in Figure 4.6.8.

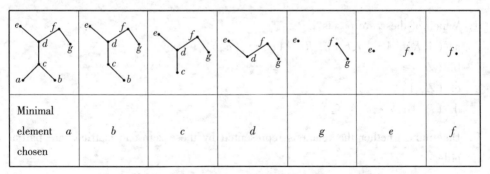

**Figure 4.6.8  Sorting process in EXAMPLE 4.6.18**

In the $k$ th step of Algorithm 4.6.1, $a_k$ is selected as a minimal element. Since there may be several different minimal elements, it implies that the selection does not need to be unique and that we can get several different total orders containing $R$.

# WORDS AND EXPRESSIONS

partial order                    偏序

| partially ordered set（或 poset） | 偏序集 |
| comparable | 可比较的 |
| total order | 全序 |
| totally ordered set | 全序集 |
| chain | 链 |
| well-ordered set | 良序集 |
| well-ordered property | 良序性 |
| lexicographic order | 字典序 |
| Hasse diagram | 哈斯图 |
| maximal element | 极大元 |
| minimal element | 极小元 |
| greatest element | 最大元 |
| least element | 最小元 |
| upper bound | 上界 |
| lower bound | 下界 |
| lattice | 格 |

## EXERCISES 4.6

1. Which of these are posets?
    a. $(A, R)$, $A = \{a, b, c\}$, $R = \{(a,a), (a,b), (b,a)\}$
    b. $(\mathbf{Z}, \geqslant)$
    c. $(\mathbf{Z}, |)$
    d. $(P(\{a,b,c,d\}), \subseteq)$

2. Determine whether the relations represented by these zero-one matrices are partial orders.

    a. $\begin{pmatrix} 1 & 1 & 0 \\ 0 & 0 & 1 \\ 1 & 0 & 1 \end{pmatrix}$  b. $\begin{pmatrix} 1 & 0 & 0 \\ 0 & 1 & 0 \\ 0 & 0 & 1 \end{pmatrix}$  c. $\begin{pmatrix} 1 & 0 & 1 & 0 \\ 0 & 1 & 1 & 0 \\ 0 & 0 & 1 & 1 \\ 1 & 1 & 0 & 1 \end{pmatrix}$

3. Determine whether the relations represented by the following directed graphs are partial orders.

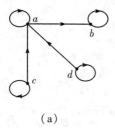
(a)
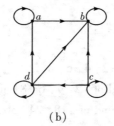
(b)
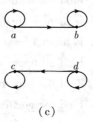
(c)

4. Find the lexicographic order for each of the pairs of these $n$-tuples
   a. $(1, 1, 1, 2, 3)$, $(1, 2, 1, 1, 3)$
   b. $(1, 0, 2, 0, 1)$, $(0, 1, 0, 1, 0)$
   c. $(1, 2, 3, 4, 5)$, $(1, 2, 3, 4, 5)$

5. Draw the Hasse diagram for divisibility on each of the following sets
   a. $\{1, 2, 3, 4, 5, 6, 7, 8\}$
   b. $\{2, 4, 6, 8, 10\}$
   c. $\{1, 2, 3, 4, 6, 12, 24, 48\}$
   d. $\{2, 4, 8, 16, 32, 64\}$

6. List all ordered pairs in the partial orderings with the accompanying Hasse diagrams.

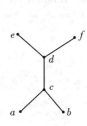
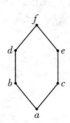
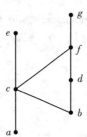

Let $(S, \leq)$ be a poset. We say that an element $b \in S$ **covers** an element $a \in S$ if $a < b$ and there is no element $c \in S$ such that $a < c < b$. The set of pairs $(a, b)$ such that $b$ covers $a$ is called the covering relation of $(S, \leq)$.

7. What is the covering relation of the poset $(P(S), \subseteq)$, where $S = \{a, b, c, d\}$?

8. Show that a finite poset can be reconstructed from its covering relation.

9. Assume the poset is $(\{2, 4, 6, 8, 9, 12, 18, 24, 27, 36, 48\}, \mid)$.
   a. Find the maximal element.

b. Find the minimal element.
c. Is there a greatest element?
d. Is there a least element?
e. Find all upper bounds of $\{2, 4, 6, 8\}$.
f. Find the least upper bound of $\{2, 4, 6, 8\}$ if it exists.
g. Find the greatest lower bound of $\{18, 24, 36, 48\}$ if it exists.

10. Show that lexicographic order is a partial order on the Cartesian product of two posets.

11. Determine whether the posets with these Hasse diagrams are lattices:

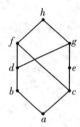 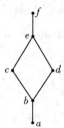

12. Show that every totally ordered set is a lattice.

13. Topologically sort the poset $(A, R)$, where $R$ is the divisibility relation on the set $A = \{1, 2, 3, 4, 6, 8, 12, 24, 36\}$.

# 5
# Graphs

Graph theory is an old subject. In 1736, the great Swiss mathematician Leonhard Euler used graph to solve the famous Königsberg bridge problem. The town of Königsberg was divided into four sections by the branches of the Pregel River. These regions were connected by seven bridges. Figure 5.0.1 depicts the regions indicated by $a,b,c,d$ and the bridges. The people in the town wondered whether it was possible to start at some location in the town, travel across each of the seven bridges exactly once, and return to the starting point.

**Figure 5.0.1   Königsberg bridge problem**

Euler's solution maybe was the first use of graph theory. We will discuss this problem further later in this chapter. Since graph theory had developed into an extensive and popular branch of mathematics, it has been used to solve problems in many fields.

## 5.1   Graph Terminology

In this section we will introduce the basic ideas of the graph theory.

**DEFINITION 5.1.1**   A pair $G = (V, E)$ is called a **graph** on $V$, where $V$ is a nonempty set of **vertices**, and $E$ is the set of **edges**. The vertices are called **nodes** or **points**, and the edges are called **links** or **lines**.

Usually we use the symbols $v(G)$ and $\varepsilon(G)$ to denote the numbers of vertices and edges in graph $G$, respectively.

Graphs can be represented graphically, and it is this graphical representation which helps us to understand many of the properties of graphs. In graphical representation, each vertex is indicated by a point, and each edge by a line joining the points which represent the ends of the edge. If an edge $e$ joins vertices $u$ and $v$, edge $e$ can be written as $uv$.

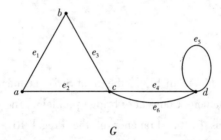

$G$

**Figure 5.1.1  A graphical representation of $G$**

A diagram of a graph $G$ is shown in Figure 5.1.1 where $G = (V, E)$, $V = \{a, b, c, d\}$, and $E = \{e_1, e_2, e_3, e_4, e_5, e_6\}$.

The types of graphs and most of the definitions and concepts in graph theory are suggested by the graphical representation. The ends of an edge are said to be **incident with** the edge, and vice versa. Two vertices which are incident with a common edge are **adjacent** (or are **neighbors**). An edge with identical ends is called a **loop**. For example, in Figure 5.1.1, $e_5$ is a loop of $G$; Vertices $a$ and $b$ are adjacent; Vertices $a$ and $c$ are incident with edge $e_2$. Edges with common ends are called **multiple** (**parallel**) **edges**. In Figure 5.1.1, $e_4$ and $e_6$ are multiple edges.

A graph is **finite** if both its vertex set and edge set are finite. A graph with just one vertex is **trivial**. A graph is **nontrivial** if it is not trivial.

**DEFINITION 5.1.2**  A graph is called to be **simple** if it has no loops and no multiple edges.

For example, graph $G$ shown in Figure 5.1.2 is a simple graph, where $G = (V, E)$, $V = \{a, b, c, d\}$, $E = \{ab, bc, ac, cd\}$.

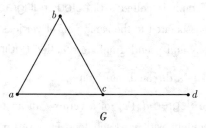

**Figure 5.1.2** A simple graph $G$

**DEFINITION 5.1.3** A graph is called a **multigraph** if it admits multiple edges. If a multigraph contains some loops, we call it a **pseudograph**.

In Figure 5.1.3, $G$ is a multigraph, and $H$ is a pseudograph.

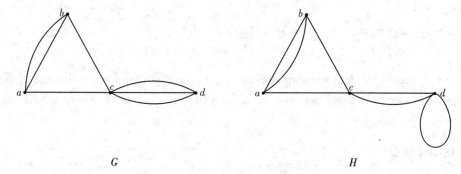

**Figure 5.1.3** Multigraph and pseudograph

**DEFINITION 5.1.4** A **directed graph** or **digraph** $G = (V, E)$ consists of a set of vertices $V$ and a set of edges $E$ that are ordered pairs of the elements of $V$.

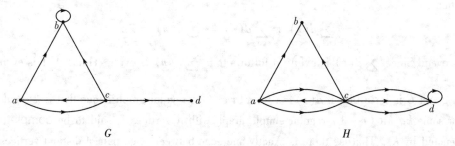

**Figure 5.1.4** Digraphs

We use an arrow pointing from $u$ to $v$ to indicate the direction of the directed edge $uv$ (or, arc $uv$). A digraph $G$ is shown in Figure 5.1.4.

**DEFINITION 5.1.5** A graph is called a **directed multigraph** if it contains directed multiple edges which are associated to the same pair of vertices with the same direction. Simple graphs, multigraphs and pseudographs are called **undirected graphs**.

Graph $H$ in Figure 5.1.4 is a directed multigraph.

**DEFINITION 5.1.6** The **degree** $d_G(v)$ of a vertex $v$ in an undirected graph $G$ is the number of edges of $G$ incident with $v$. Each loop at a vertex contributes twice to the degree of that vertex. If $d_G(v) = 0$, $v$ is said to be **isolated**. If $d_G(v) = 1$, then $v$ is called a **leaf** (or **pendent vertex**, or **pendant vertex**). If there is no ambiguous, we use $d(v)$ instead of $d_G(v)$.

**EXAMPLE 5.1.1** What are the degrees of the vertices in graph $G$ and $H$ displayed in Figure 5.1.3?
*Solution*: $d_G(a) = 3 \quad d_G(b) = 3 \quad d_G(c) = 5 \quad d_G(d) = 3$
$\qquad d_H(a) = 3 \quad d_H(b) = 3 \quad d_H(c) = 4 \quad d_H(d) = 4$

Since an edge is incident with two vertices, each edge contributes two to the sum of the degrees of vertices. Each loop counts two at that vertex as well. We then have the following theorem.

**THEOREM 5.1.1** $\sum_{v \in V} d(v) = 2\varepsilon$

**THEOREM 5.1.2** In any undirected graph, the number of vertices with odd degree is even.
*Proof*: Let $V_1$ and $V_2$ be the sets of vertices of odd and even degrees in $G$, respectively. Then

$$\sum_{v \in V_1} d(v) + \sum_{v \in V_2} d(v) = \sum_{v \in V} d(v) = 2\varepsilon$$

is even. Since $\sum_{v \in V_2} d(v)$ is even, it follows that $\sum_{v \in V_1} d(v)$ is even. Thus $|V_1|$ is even.

A graph $G$ is $k$-**regular** if $d(v) = k$ for all $v \in V$. A regular graph is one that is $k$-regular for some $k$. An $(n-1)$-regular simple graph with $n$ vertices is said to be **complete**, denoted by $K_n$. That is, there is exactly one edge between each pair of distinct vertices. The graphs $K_n$, for $n = 1,2,3,4,5,6$ are displayed in Figure 5.1.5.

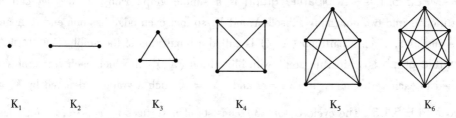

Figure 5.1.5  The graphs $K_n$ for $n = 1,2,3,4,5,6$

**EXAMPLE 5.1.2**  How many edges are there in a 5-regular graph with ten vertices?

*Solution*: By THEOREM 5.1.1, $\sum_{v \in V} d(v) = 2\varepsilon$. It follows that $2\varepsilon = 5 \times 10$. Therefore $\varepsilon = 25$.

**DEFINITION 5.1.7**  If $uv$ is a directed edge of digraph $G$, vertex $u$ is called the **initial vertex** of $uv$, and $v$ is called the **terminal vertex** of $uv$. The initial vertex and terminal vertex of a loop are the same vertex.

**DEFINITION 5.1.8**  The **in-degree** $d_G^-(v)$ of a vertex $v$ in digraph $G$ is the number of edges with $v$ as their terminal vertex. The **out-degree** $d_G^+(v)$ is the number of edges with $v$ as their initial vertex. (Note that a loop at a vertex contributes 1 to both the in-degree and the out-degree of this vertex.)

**EXAMPLE 5.1.3**  Find the in-degree and out-degree of each vertex in the graph $H$ shown in Figure 5.1.4. Compute $\sum_{v \in V} d_H^-(v)$ and $\sum_{v \in V} d_H^+(v)$ and count the number of edges.

*Solution*:  $d_H^-(a) = 1 \quad d_H^+(a) = 3 \quad d_H^-(b) = 2 \quad d_H^+(b) = 0$
$d_H^-(c) = 2 \quad d_H^+(c) = 5 \quad d_H^-(d) = 4 \quad d_H^+(d) = 1$

Thus, $\sum_{v \in V} d_H^-(v) = 9$, $\sum_{v \in V} d_H^+(v) = 9$, and $\varepsilon = 9$.

In EXAMPLE 5.1.3, the sum of the in-degrees and the sum of the out-degrees of all vertices are the same. Furthermore, these sums are equal to the number of edges in the graph. In fact, generally we have the following theorem.

**THEOREM 5.1.3**  Let $G = (V, E)$ be a graph with directed edges. Then

$$\sum_{v \in V} d_G^-(v) = \sum_{v \in V} d_G^+(v) = \varepsilon$$

We conclude this section by introducing some special classes of graphs.

**EXAMPLE 5.1.4** A **bipartite graph** is a simple graph whose vertex set can be partitioned into two nonempty subsets $X$ and $Y$, so that each edge has one end in $X$ and one end in $Y$. Such a partition $(X,Y)$ is called a bipartition of the graph. A **complete bipartite graph** is a bipartite graph with bipartition $(X,Y)$ in which each vertex of $X$ is joined to each vertex of $Y$. If $|X|=m$ and $|Y|=n$, such a graph is denoted by $K_{m,n}$.

**EXAMPLE 5.1.5** The cycle $C_n(n \geqslant 3)$ consists of $n$ vertices $v_1, v_2, \cdots, v_n$ and edges $v_1v_2, v_2v_3, \cdots, v_{n-1}v_n, v_nv_1$.

The graphs displayed in Figure 5.1.6 are: (a) a bipartite graph; (b) the complete bipartite graph $K_{2,3}$; (c) the cycle $C_6$; (d) $K_{1,8}$. $K_{1,n}$ is called a star, in which all vertices are connected to the center vertex. From Figure 5.1.6(d) it is obvious that $K_{1,8}$ is a star.

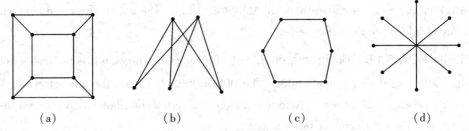

(a)　　　　　(b)　　　　　(c)　　　　　(d)

**Figure 5.1.6　Bipartite graphs**

## WORDS AND EXPRESSIONS

| | |
|---|---|
| vertex | 点 |
| link | 边,连接 |
| incident | 关联的 |
| adjacent | 邻接的 |
| parallel | 平行的 |
| multiple edges | 重边 |
| simple graph | 简单图 |
| pseudograph | 伪图 |
| degree of vertex | 点度 |
| isolated vertex | 孤立点 |
| pendent vertex | 悬挂点 |

| | |
|---|---|
| leaf | 叶子 |
| regular graph | 正则图 |
| complete graph | 完全图 |
| in-degree | 入度 |
| out-degree | 出度 |
| bipartite graph | 二部图 |
| bipartition | 二划分 |
| complete bipartite graph | 完全二部图 |
| cycle | 圈 |
| star | 星图 |

## EXERCISES 5.1

1. Determine whether each of the graphs shown below is a simple graph, a multi-graph (but not a simple graph), a pseudograph (but not a multi-graph), a directed graph, or a directed multi-graph (but not a directed graph).

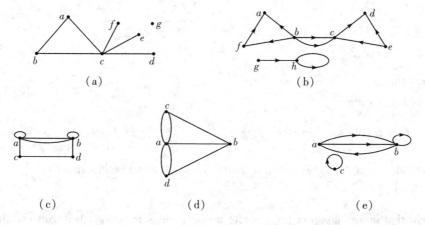

Computer programs can be executed more rapidly when certain statements in the program are executed concurrently. But we cannot execute a statement that needs results from other statements that have not yet been executed. The dependence of some statements on previous statements can be represented by a digraph, in which we represent each statement by a vertex, and use a directed edge $uv$ to indicate that statement $v$ cannot be executed before $u$ completed. The resulting graph is called a **precedence graph**.

2. Construct a precedence graph for the following program:
   (S1) $a := 2$
   (S2) $b := a + 2$
   (S3) $c := 1$
   (S4) $d := a * c + 3$
   (S5) $e := d - 2$
   (S6) $f := c + e$
   (S7) $g := b - f$

3. Which statements must be executed before (S6) is executed in the program above?

4. How many edges does a graph have if it has vertices of degree 4, 3, 3, 3, 2, 1? Draw such a graph.

5. Does there exist a simple graph with six vertexes of the following degrees? If so, draw such a graph.
   a. 1,2,3,4,5,6
   b. 0,1,2,3,4,5
   c. 1,1,1,1,1,1
   d. 3,3,3,3,2,2

6. Show that
   a. $\varepsilon(K_{m,n}) = mn$
   b. If $G$ is simple and bipartite, then $\varepsilon \leq \dfrac{v^2}{4}$.

7. Show that if a $k$-regular bipartite graph where $k > 0$ has bipartition $(X, Y)$ then $|X| = |Y|$.

8. Show that in any group of two or more people, there are always two with exactly the same number of friends inside the group.

9. The $n$-cube is the graph whose vertices are the ordered $n$-tuples of 0's and 1's. Two vertices are joined if and only if they differ in exactly one coordinate. Show that the $n$-cube has $2^n$ vertices, $n \cdot 2^{n-1}$ edges and it is a bipartite graph.

10. The complement $\overline{G}$ of a simple graph $G = (V, E)$ is the simple graph with vertex set $V$, and two vertices are adjacent in $\overline{G}$ if and only if they are not adjacent in $G$.

Describe the graphs $\overline{K}_n$ and $\overline{K}_{m,n}$.

## 5.2 Representing Graphs and Graph Isomorphism

There are many ways to represent graphs. In order to simplify computation, graphs can be represented with matrices. Two types of matrices commonly used to represent graphs are adjacency matrix and incidence matrix.

Suppose that $G = (V, E)$ is an undirected graph and $|V| = n$. Let us denote the vertices of $G$ by $v_1, v_2, \cdots, v_n$. The **adjacency matrix** of $G$ is the $n \times n$ matrix $A = [a_{ij}]$ (or $A_G = [a_{ij}]$), in which $a_{ij}$ is the number of edges joining $v_i$ and $v_j$. A graph $G$ and its adjacency matrix are shown in Figure 5.2.1.

Note that:
1. An adjacency matrix of a graph is based on the ordering chosen for the vertices, i.e., different orders of vertice correspond to different adjacency matrices.
2. The adjacent matrix of an undirected graph is symmetric, since $a_{ij} = a_{ji}$, $i, j = 1, 2, \cdots, n$.

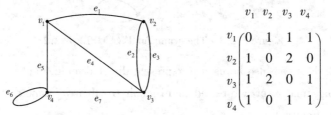

**Figure 5.2.1  Graph $G$ and its adjacency matrix**

**EXAMPLE 5.2.1** Use an adjacency matrix to represent the graph shown in Figure 5.2.2

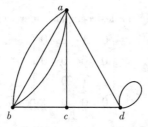

**Figure 5.2.2  The graph for EXAMPLE 5.2.1**

*Solution:* We order the vertices as $a, b, c, d$. The adjacency matrix is

$$\begin{pmatrix} 0 & 3 & 1 & 1 \\ 3 & 0 & 1 & 0 \\ 1 & 1 & 0 & 1 \\ 1 & 0 & 1 & 1 \end{pmatrix}$$

**EXAMPLE 5.2.2** Draw a graph with the following adjacency matrix with respect to the ordering of the vertices $a, b, c, d$.

$$\begin{pmatrix} 1 & 0 & 1 & 1 \\ 0 & 0 & 0 & 1 \\ 1 & 0 & 1 & 0 \\ 1 & 1 & 0 & 0 \end{pmatrix}$$

*Solution*: A graph with this adjacency matrix is shown in Figure 5.2.3.

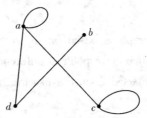

**Figure 5.2.3  The graph for EXAMPLE 5.2.2**

Adjacency matrices can also be used to represent directed multi-graphs. Let $A_G = [a_{ij}]$ be an adjacency matrix of a directed multi-graph $G$, then $a_{ij}$ denotes the number of directed edges from $v_i$ to $v_j$.

**EXAMPLE 5.2.3** Use an adjacency matrix to represent the directed multi-graph $H$ shown in Figure 5.1.4.

*Solution*: We order the vertices in the lexicographic order: $a, b, c, d$. The adjacency matrix is

$$\begin{pmatrix} 0 & 1 & 2 & 0 \\ 0 & 0 & 0 & 0 \\ 1 & 1 & 0 & 3 \\ 0 & 0 & 0 & 1 \end{pmatrix}$$

Another kind of matrix associated with graphs is incidence matrix. Let $G = (V, E)$ be an undirected graph. Let $v_1, v_2, \cdots, v_n$ be the vertices of $G$ and $e_1, e_2, \cdots, e_m$ be the edges of $G$. Then the incidence matrix of $G$ is the matrix $M = [m_{ij}]$ (or $M_G = [m_{ij}]$), where

$m_{ij}$ is the number of times $(0,1,\text{or }2)$ that $v_i$ and $e_j$ are incident. For example, the graph G shown in Figure 5.2.1 can be represented by the following incidence matrix:

$$\begin{array}{c} \\ v_1 \\ v_2 \\ v_3 \\ v_4 \end{array} \begin{array}{c} \begin{array}{ccccccc} e_1 & e_2 & e_3 & e_4 & e_5 & e_6 & e_7 \end{array} \\ \left( \begin{array}{ccccccc} 1 & 0 & 0 & 1 & 1 & 0 & 0 \\ 1 & 1 & 1 & 0 & 0 & 0 & 0 \\ 0 & 1 & 1 & 1 & 0 & 0 & 1 \\ 0 & 0 & 0 & 0 & 1 & 2 & 1 \end{array} \right) \end{array}$$

Note that:
1. Multiple edges are represented in the incident matrix using columns with identical entries, since these edges are incident with the same pair of vertices;
2. Loops are represented using a column with exactly one entry equal to 2, corresponding to the vertex that is incident with this loop.

Now, let us look at the graphs G and H shown in Figure 5.2.4.

Figure 5.2.4  Two graphs having same structure

If we rearrange the positions of vertices of H as follows:

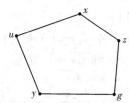

then we find graphs G and H have the same structure. In many applications, we often need to know whether it is possible to draw two graphs in the same way. The following definition is a useful concept for graphs with the same structure.

**DEFINITION 5.2.1**  If graph $G = (V_1, E_1)$ and $H = (V_2, E_2)$ are simple, and there is a bijection $\theta: V_1 \to V_2$ such that $ab \in E_1$ if and only if $\theta(a)\theta(b) \in E_2$, then $G$ and $H$ are **isomorphic**. Bijection $\theta$ is an **isomorphism** between $G$ and $H$.

In other words, when two simple graphs are isomorphic, there is a bijection between the vertices of the graphs that preserves the adjacency relationship.

To show that two graphs are isomorphic, one must indicate an isomorphism between them. The mapping $\theta$ defined by

$$\theta(a) = x \quad \theta(b) = u \quad \theta(c) = y \quad \theta(d) = v \quad \theta(e) = z$$

is an isomorphism between graphs $G$ and $H$ displayed in Figure 5.2.4. $G$ and $H$ clearly have the same structure, and differ only in the names of the vertices. Since it is the structural properties that we are primarily interested in, we shall omit labels when drawing graphs. An unlabelled graph can be thought as a representative of an equivalence class of isomorphic graphs. We assign labels to vertices and edges in a graph mainly for the purpose of referring to them.

**EXAMPLE 5.2.4** Show that the graphs $G = (V_1, E_1)$ and $H = (V_2, E_2)$ displayed in Figure 5.2.5 are isomorphic.

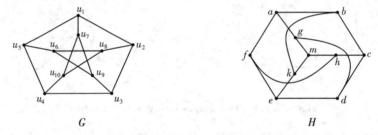

G　　　　　　　　　　　　　　H

**Figure 5.2.5　Isomorphic graphs for EXAMPLE 5.2.4**

*Solution* : The bijection $\theta$ defined by

$\theta(u_1) = a \quad \theta(u_2) = f \quad \theta(u_3) = e \quad \theta(u_4) = k \quad \theta(u_5) = b$
$\theta(u_6) = c \quad \theta(u_7) = g \quad \theta(u_8) = h \quad \theta(u_9) = d \quad \theta(u_{10}) = m$

preserves all adjacencies.

Since an isomorphism between two graphs is a bijection, isomorphic graphs must have the same number of vertices and the same number of edges. An isomorphism preserves adjacencies. Therefore it preserves graph substructures. Consequently isomorphic graphs have the same degree sequence of vertices. These properties are called **invariants** with respect to isomorphism of graphs. It is often difficult to determine whether two graphs are isomorphic, but if any of these invariants in two graphs differ, the graphs cannot be isomorphic. For example, the following two graphs $G$ and $F$ can be used to model saturated hydrocarbon compounds $C_4H_{10}$.

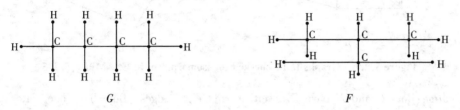

They share the same number of vertices, the same number of edges, and the same degree sequence. In structure $G$, the four carbons of degree 4 form a path:

But in structure $F$, the four carbons of degree 4 form a different substructure:

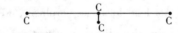

Therefore, $G$ and $F$ are not isomorphic.

The number of vertices, the number of edges, and the degree sequence of vertices are all invariants under isomorphism. But these invariants are not enough to determine whether two graphs are isomorphic. Unfortunately, there are no useful sets of invariants known that can be used to determine whether graphs are isomorphic.

Suppose that we have established a mapping $\theta$ from the vertex set of graph $G$ to the vertex set of graph $H$. To show that $\theta$ is an isomorphism, we need to prove that $\theta$ preserves edges. To achieve this, we can show that the adjacency matrix of $G$ is the same as the adjacency matrix of $H$, when the rows and columns of the second matrix are labeled to correspond to the images under $\theta$ of the vertices of $G$ that are the labels of these rows and columns in the first matrix. But if $\theta$ turned out not to be an isomorphism, we could not claim that $G$ and $H$ are not isomorphic, since another correspondence of the vertices in $G$ and $H$ may be an isomorphism.

**EXAMPLE 5.2.5** Determine whether the graphs $G$ and $H$ displayed in Figure 5.2.6 are isomorphic.

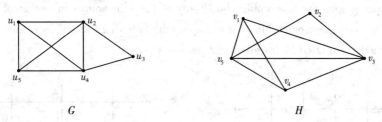

**Figure 5.2.6** Graphs to be checked for isomorphism in EXAMPLE 5.2.5

*Solution:* Both $G$ and $H$ have five vertices and eight edges. Both have one vertex of degree two and two vertices of degree three and two vertices of degree four. Since $G$ and $H$ agree with respect to these invariants, it is reasonable to try to find an isomorphism $\theta$.

We now define a mapping $\theta$ and then determine whether it is an isomorphism. Since $d(u_3) = 2$ and it is the only vertex of degree two, the image of $u_3$ must be $v_2$. Since $u_2$ is adjacent to $u_3$, the possible image of $u_2$ is $v_3$ or $v_5$. We arbitrarily set $\theta(u_2) = v_3$. (If we found that this choice did not lead to an isomorphism, we could then try $\theta(u_2) = v_5$.) Continuing in this way, using the adjacency of vertices and degrees as a guide, we set $\theta(u_4) = v_5, \theta(u_1) = v_1, \theta(u_5) = v_4$. To see whether $\theta$ preserves edges, we examine the adjacency matrix of $G$:

$$A_G = \begin{array}{c} \\ u_1 \\ u_2 \\ u_3 \\ u_4 \\ u_5 \end{array} \begin{array}{c} u_1\ u_2\ u_3\ u_4\ u_5 \\ \begin{pmatrix} 0 & 1 & 0 & 1 & 1 \\ 1 & 0 & 1 & 1 & 1 \\ 0 & 1 & 0 & 1 & 0 \\ 1 & 1 & 1 & 0 & 1 \\ 1 & 1 & 0 & 1 & 0 \end{pmatrix} \end{array}$$

and the adjacency matrix of $H$ with the rows and columns labelled by the images of the corresponding vertices in $G$:

$$A_H = \begin{array}{c} \\ v_1 \\ v_3 \\ v_2 \\ v_5 \\ v_4 \end{array} \begin{array}{c} v_1\ v_3\ v_2\ v_5\ v_4 \\ \begin{pmatrix} 0 & 1 & 0 & 1 & 1 \\ 1 & 0 & 1 & 1 & 1 \\ 0 & 1 & 0 & 1 & 0 \\ 1 & 1 & 1 & 0 & 1 \\ 1 & 1 & 0 & 1 & 0 \end{pmatrix} \end{array}$$

Since $A_G = A_H$, it follows that $\theta$ preserves edges. We conclude that $\theta$ is an isomorphism, so that $G$ and $H$ are isomorphic.

## WORDS AND EXPRESSIONS

| | |
|---|---|
| adjacency matrix | 邻接矩阵 |
| incidence matrix | 关联矩阵 |
| isomorphism | 同构 |
| invariant with respect to isomorphism | 同构不变量 |

## EXERCISES 5.2

1. Use adjacency matrices to represent the following graphs.

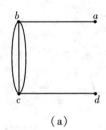
(a)
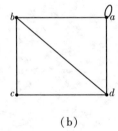
(b)
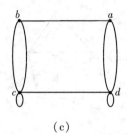
(c)

2. Draw graphs for the given adjacency matrices below.

$$\begin{pmatrix} 0 & 0 & 1 \\ 1 & 0 & 1 \\ 1 & 1 & 0 \end{pmatrix} \quad \begin{pmatrix} 0 & 0 & 2 & 1 \\ 0 & 0 & 1 & 0 \\ 2 & 1 & 0 & 1 \\ 1 & 0 & 1 & 0 \end{pmatrix} \quad \begin{pmatrix} 1 & 1 & 0 & 0 \\ 1 & 0 & 2 & 0 \\ 0 & 2 & 0 & 1 \\ 0 & 0 & 1 & 1 \end{pmatrix}$$

(a) (b) (c)

3. Find the adjacency matrices for the given directed multigraphs below.

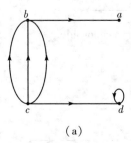
(a)
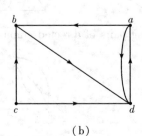
(b)
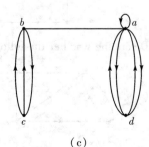
(c)

4. Use incidence matrices to represent the graphs in Exercises 1 and 2.

5. Let $M$ and $A$ be the incidence matrix and the adjacency matrix of a simple graph $G$, respectively.

   a. Show that every column sum of $M$ is 2.

   b. What are the column sums of $A$?

6. Determine whether the following pairs of graphs are isomorphic. Exhibit an isomorphism if yes, or provide a rigorous argument if none exists.

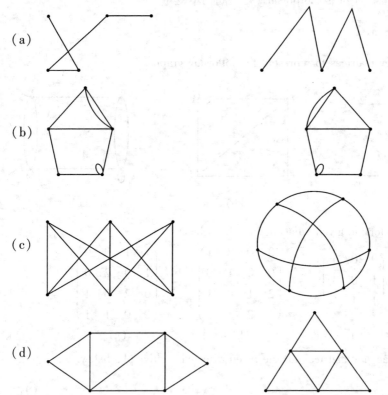

7. Determine whether the following pairs of directed graphs are isomorphic.

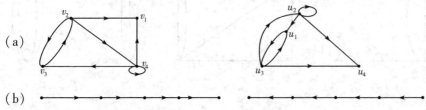

(c)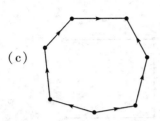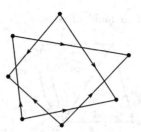

8. Show that the vertices of a bipartite graph with two or more vertices can be enumerated so that the adjacency matrix of the graph has the form
$$\begin{pmatrix} 0 & A_{12} \\ \hline A_{21} & 0 \end{pmatrix}$$
where $A_{21}$ is the transpose of $A_{12}$.

9. Find an adjacency matrix for each of the following graphs.
$$K_n \qquad C_n \qquad K_{m,n}$$

A simple graph $G$ is called **self-complementary** if $G$ and $\overline{G}$ are isomorphic.

10. Find a self-complementary simple graph with five vertices.

11. Show that if $G$ is a self-complementary simple graph with $n$ vertices, then
$$n \equiv 0 \text{ or } 1 \pmod{4}.$$

## 5.3 Subgraphs

**DEFINITION 5.3.1** Let $G = (V_1, E_1)$ and $H = (V_2, E_2)$ be graphs. If $V_2 \subseteq V_1$ and $E_2 \subseteq E_1$, then $H$ is called a **subgraph** of $G$, denoted by $H \subseteq G$. If $H \subseteq G$ and $H \neq G$, then $H$ is called a **proper subgraph** of $G$. If $H \subseteq G$ and $V_1 = V_2$, then $H$ is called a **spanning subgraph** of $G$.

Figure 5.3.1 displays a graph and its spanning subgraph.

From the definition, a graph is a spanning subgraph of itself.

**DEFINITION 5.3.2** Suppose that $V'$ is a nonempty subset of $V$. The subgraph of $G$ induced by $V'$, denoted by $G[V']$, is the graph whose vertex set is $V'$ and edge set is the set of those edges of $G$ that have both ends in $V'$. The induced subgraph $G[V \backslash V']$ is denoted by $G-V'$. It is the subgraph obtained from $G$ by removing the vertices in $V'$

together with their incident edges.

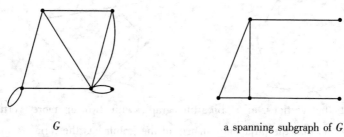

G　　　　　　　　　　　　　　a spanning subgraph of G

**Figure 5.3.1　A graph and its spanning subgraph**

**DEFINITION 5.3.3** Suppose that $E'$ is a nonempty subset of $E$. The subgraph of $G$, whose vertex set is the set of ends of edges in $E'$ and whose edge set is $E'$, is called the subgraph of $G$ induced by $E'$ and is denoted by $G[E']$. $G[E']$ is an **edge-induced subgraph** of $G$. The spanning subgraph of $G$ with edge set $E \setminus E'$ is written as $G-E'$. It is the subgraph obtained from $G$ by deleting the edges in $E'$.

We can also add some edges into a graph. $G + E_0$ is a graph obtained by adding a set of edges $E_0$ into $G$. Some types of subgraphs are displayed in Figure 5.3.2.

Two or more graphs can be combined in various ways as described in the following definitions.

**DEFINITION 5.3.4** The **union** of two simple graphs $G_1 = (V_1, E_1)$ and $G_2 = (V_2, E_2)$, denoted by $G_1 \cup G_2$, is the simple graph with vertex set $V_1 \cup V_2$ and edge set $E_1 \cup E_2$.

If $V_1 \cap V_2 \neq \emptyset$, we can define the intersection of two graphs in the following way.

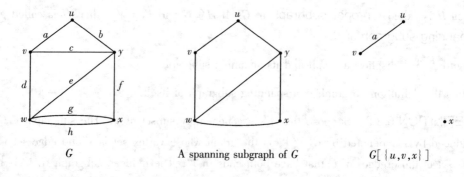

G　　　　　　　　A spanning subgraph of G　　　　　　　$G[\{u,v,x\}]$

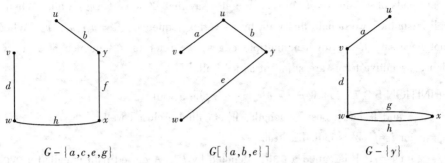

Figure 5.3.2 Spanning subgraph, induced subgraph, and edge-induced subgraph of $G$

**DEFINITION 5.3.5** Let $G_1 = (V_1, E_1)$ and $G_2 = (V_2, E_2)$ be two simple graphs and $V_1 \cap V_2 \neq \emptyset$. The **intersection** of $G_1$ and $G_2$, denoted by $G_1 \cap G_2$, is the simple graph with vertex set $V_1 \cap V_2$ and edge set $E_1 \cap E_2$.

Figure 5.3.3 displays two graphs and their union and intersection.

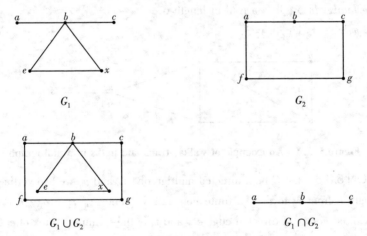

Figure 5.3.3 Union and intersection of graphs

Many problems can be modeled by some special subgraphs. For instance, problems of efficiently planning routes for mail delivery, message transmitting in computer networks and so on, can be solved using models that involve paths in graphs.

**DEFINITION 5.3.6** Let $n$ be a nonnegative integer and $G$ an undirected graph. $v_0$ and $v_n$ are two vertices of $G$. A **walk of length** $n$ from $v_0$ to $v_n$ is a finite sequence $W = v_0 e_1 v_1 e_2 \cdots e_n v_n$, whose terms are alternatively vertices and edges of $G$, such that $v_i$ and

$v_{i+1}$ are ends of $e_{i+1}$ for $0 \leq i \leq n-1$. We also say that $W$ is a $(v_0, v_n)$-walk. When the graph is simple, we denote this walk by its vertex sequence: $W = v_0 v_1 v_2 \cdots v_n$. When it is not necessary to distinguish multiple edges, we refer to a sequence of vertices in which consecutive terms are adjacent as a "walk".

**DEFINITION 5.3.7** Let $W = v_0 e_1 v_1 e_2 v_2 \cdots e_n v_n$ be a walk.
If $v_0 = v_n$, and $W$ has a positive length, $W$ is called a closed walk.
If $e_i \neq e_j$ for $i \neq j$, $W$ is called a **trail**.
If $v_i \neq v_j$ for $i \neq j$, $W$ is called a **path**, denoted by $P_n$. A closed path is called a **cycle**, denoted by $C_n$. If $n$ is even, we call $C_n$ an even cycle, and if $n$ is odd, we call $C_n$ an odd cycle.

**EXAMPLE 5.3.1** Figure 5.3.4 illustrates a walk, a trail, and a path in a multigraph, where we can find:
A walk: $uavfzguavbw$; this walk is of length 5.
A trail: $ugzhuaveydwcx$; it is a trail of length 6.
A path: $xcwdyiz$; its length is 3.

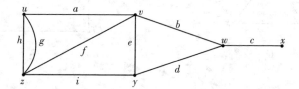

**Figure 5.3.4** An example of walks, trails and paths in a multigraph

**DEFINITION 5.3.8** Let $G$ be a directed multigraph. $v_0$ and $v_n$ are two vertices of $G$. A path of length $n$ from $v_0$ to $v_n$ is a finite non-null sequence $v_0 e_1 v_1 e_2 v_2 \cdots e_n v_n$ such that $v_{i-1}$ is the initial vertex of directed edge $e_i$, and $v_i$ is the terminal vertex of $e_i$ for $1 \leq i \leq n-1$. When there is no parallel edges in the directed graph, this path is denoted by its vertex sequence $v_0 v_1 v_2 \cdots v_n$. A path of a directed multigraph sometimes is called a directed path. A closed directed path with length at least three is called a directed cycle.

In an undirected graph $G$, two vertices $u$ and $v$ are **connected** if there is an $(u,v)$-path in $G$. If every pair of distinct vertices of $G$ are connected, $G$ is called to be **connected**. Connection is an equivalence relation on the vertex set $V$. Thus, there is a partition of $V$ into nonempty subsets $V_1, V_2, \cdots, V_k$ such that two vertices $u$ and $v$ are connected if and only if both $u$ and $v$ belong to the same set $V_i$. The induced subgraphs $G[V_1], G[V_2], \cdots$, $G[V_k]$ are called the **components** of $G$. If $G$ has more than one component, then $G$ is called to be **disconnected**. We use the symbol $w(G)$ to denote the number of components of $G$.

**EXAMPLE 5.3.2** The graph $G$ shown in Figure 5.3.5 has three components $G[\{a,b,c\}]$, $G[\{d,e\}]$, and $G[\{f\}]$.

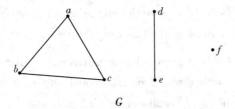

$G$

**Figure 5.3.5** A graph with 3 components

In a directed graph $G$, if for any two vertices $u$ and $v$, there is a path from $u$ to $v$ and there is also a path from $v$ to $u$, then $G$ is called to be **strongly connected**. If the underlying undirected graph of $G$ is connected, $G$ is called to be **weakly connected**.

**EXAMPLE 5.3.3** Determine whether graphs $G$ and $H$ shown in Figure 5.3.6 are strongly connected, or weakly connected.

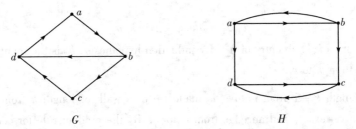

$G$ $H$

**Figure 5.3.6** Graphs for EXAMPLE 5.3.3

*Solution*: $G$ is strongly connected because there is a path between any two vertices. Certainly, $G$ is also weakly connected. But $H$ is not strongly connected, because there is no directed path from $c$ to $a$ in this graph. However, $H$ is weakly connected because its underlying undirected graph is connected, that is, there is a path between any two vertices in the underlying undirected graph.

Similar to the concept of component in an undirected graph, we can define strongly component in a directed graph. The subgraphs of a directed graph $G$ that are strongly connected but not contained in any larger strongly connected subgraphs (so they are maximal strongly connected subgraphs) are called **strongly components** of $G$.

The directed graph $G$ shown in Figure 5.3.6 has just one strongly component because $G$ itself is a strongly connected digraph. But $H$ has two strongly components $H[\{a,b\}]$ and $H[\{c,d\}]$.

The number of walks between two vertices in a graph can be determined using its adjacency matrix, as described in the following theorem.

**THEOREM 5.3.1** Let $G$ be a graph. $A$ is an adjacency matrix of $G$ with respect to the ordering $v_1, v_2, \cdots, v_n$ (with directed or undirected edges, multiple edges and loops allowed). The number of $(v_i, v_j)$-walks of length $k$ in $G$ where $k$ is a positive integer, is the $(i,j)$ th entry of $A^k$.

*Proof*: The theorem will be proved using mathematical induction. Firstly, the number of walks from $v_i$ to $v_j$ of length 1 is the $(i,j)$ th entry of $A$, since $a_{ij}$ is equal to the number of edges from $v_i$ to $v_j$.

Assume that the $(i,j)$ th entry of $A^k$ is the number of different walks of length $k$ from $v_i$ to $v_j$. This is the induction hypothesis. Since $A^{k+1} = A^k \cdot A$, the $(i,j)$ th entry of $A^{k+1}$ is equal to

$$b_{i1}a_{1j} + b_{i2}a_{2j} + \cdots + b_{in}a_{nj}$$

where $b_{il}$ is the $(i,l)$ th entry of $A^k$. By induction hypothesis, $b_{il}$ is the number of walks of length $k$ from $v_i$ to $v_l$.

A walk of length $k+1$ from $v_i$ to $v_j$ is made up of a walk of length $k$ from $v_i$ to some intermediate vertex $v_l$, and an edge from $v_l$ to $v_j$. By the product rule for counting, the number of such walks is the product of the number of walks of length $k$ from $v_i$ to $v_l$, namely, $b_{il}$, and the number of edges from $v_l$ to $v_j$, namely, $a_{lj}$. When these products are added for all possible intermediate vertices $v_l$, the desired result follows by the sum rule for counting.

**EXAMPLE 5.3.4** In the simple graph $G$ shown in Figure 5.3.7, find the number of walks of length 3 from $a$ to $b$.

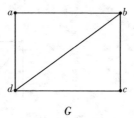

**Figure 5.3.7** The Graph for EXAMPLE 5.3.4

*Solution*: The adjacency matrix of $G$ with the ordering $a, b, c, d$ is

$$A = \begin{pmatrix} 0 & 1 & 0 & 1 \\ 1 & 0 & 1 & 1 \\ 0 & 1 & 0 & 1 \\ 1 & 1 & 1 & 0 \end{pmatrix}$$

Then $A^2 = A \cdot A = \begin{pmatrix} 0 & 1 & 0 & 1 \\ 1 & 0 & 1 & 1 \\ 0 & 1 & 0 & 1 \\ 1 & 1 & 1 & 0 \end{pmatrix} \cdot \begin{pmatrix} 0 & 1 & 0 & 1 \\ 1 & 0 & 1 & 1 \\ 0 & 1 & 0 & 1 \\ 1 & 1 & 1 & 0 \end{pmatrix} = \begin{pmatrix} 2 & 1 & 2 & 1 \\ 1 & 3 & 1 & 2 \\ 2 & 1 & 2 & 1 \\ 1 & 2 & 1 & 3 \end{pmatrix}$

$A^3 = A \cdot A \cdot A = \begin{pmatrix} 2 & 1 & 2 & 1 \\ 1 & 3 & 1 & 2 \\ 2 & 1 & 2 & 1 \\ 1 & 2 & 1 & 3 \end{pmatrix} \cdot \begin{pmatrix} 0 & 1 & 0 & 1 \\ 1 & 0 & 1 & 1 \\ 0 & 1 & 0 & 1 \\ 1 & 1 & 1 & 0 \end{pmatrix} = \begin{pmatrix} 2 & 5 & 2 & 5 \\ 5 & 4 & 5 & 5 \\ 2 & 5 & 2 & 5 \\ 5 & 5 & 5 & 4 \end{pmatrix}$

There are five walks of length 3 from $a$ to $b$, namely, $abab, adab, abdb, abcb, adcb$, in which $adcb$ is a path of length three from $a$ to $b$.

We will close this section with some further comments on isomorphism. If two graphs $G$ and $H$ are isomorphic, then their corresponding subgraphs are isomorphic. For example, suppose that $G$ and $H$ are isomorphic, if $G$ contains a path $P_k$ of length $k$, then $H$ must contain a path of length $k$ to correspond to $P_k$. If $G$ contains a cycle $C_l$ of length $l$, then $H$ must contain a cycle of length $l$ to correspond to $C_l$.

**EXAMPLE 5.3.5** Determine whether the graphs $G$ and $H$ displayed in Figure 5.3.8 are isomorphic.

*Solution*: Both $G$ and $H$ have six vertices and eight edges. They have the same degree sequence $(2,2,2,3,3,4)$. So the three invariants, which are the number of vertices, the number of edges, and the degree sequence, all agree for the two graphs. However,

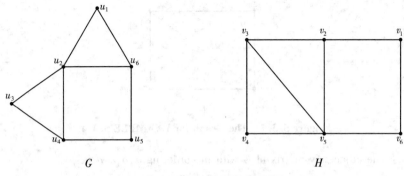

**Figure 5.3.8** Graphs for EXAMPLE 5.3.5

in $G$ the three vertices of degree two, namely $u_1, u_3, u_5$, induce a subgraph which consists of three isolated vertices. In $H$, the three vertices of degree two, namely, $v_1$, $v_4, v_6$, induce a subgraph which consists of an edge $v_1 v_6$ and an isolated vertex $v_4$. Therefore $G$ and $H$ are not isomorphic.

## WORDS AND EXPRESSIONS

| | |
|---|---|
| Subgraph | 子图 |
| Proper subgraph | 真子图 |
| Spanning subgraph | 支撑子图，生成子图 |
| Induced subgraph | 导出子图 |
| Edge-induced subgraph | 边导出子图 |
| Walk | 链 |
| Trail | 迹 |
| Path | 路 |
| Connected graph | 连通图 |
| Disconnected graph | 不连通图 |
| Component | 连通支图 |
| Strongly connected | 强连通的 |
| Weakly connected | 弱连通的 |

## EXERCISES 5.3

1. Show that every simple graph on $n$ vertices is isomorphic to a subgraph of $K_n$.

2. Show that:

a. Every induced subgraph of a complete graph is complete.

b. Every induced subgraph of a bipartite graph is bipartite.

3. Describe how $A_{G-V'}$ can be obtained from $A_G$, and how $M_{G-V'}$ and $M_{G-E'}$ can be obtained from $M_G$.

4. Does each of the following sequences of vertices form a walk in the graph below? Which walks are trail? Which are paths? What are the lengths of those that are walks?

   acdedcg    abcdfg    agcdfed    bgacdef

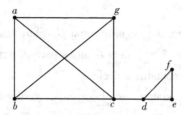

5. Does each of these sequences of vertices form a walk in the following graph? Which are trails? Which are paths? What are the lengths of those that are walks?

   adefdabcad    bcdad    fdcba    aefdad

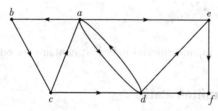

6. Determine the number of components of each of these graphs.

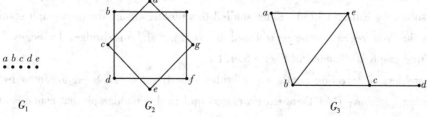

7. Find the strongly components for each of these graphs.

8. Find the number of walks of length 3 from $a$ to $c$ in graph $G_3$ of Exercise 6.

9. Find the number of walks of length 3 from $a$ to $c$ in the directed graph $G_1$ of Exercise 7.

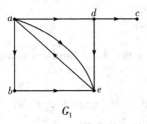
$G_1$

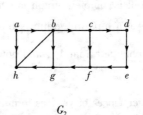
$G_2$

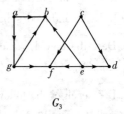
$G_3$

10. Show that in any simple graph there is a path from any vertex of odd degree to some other vertex of odd degree.

11. Let $\delta$ denote the minimum degree of a graph $G$. Show that:

    a. If $G$ is simple and $\delta \geq k$, then $G$ has a path of length $k$;

    b. If $\delta \geq 2$, then $G$ contains a cycle.

12. Show that if $G$ is a simple graph with $n$ vertices and it has more than $\frac{(n-1)(n-2)}{2}$ edges, then $G$ is connected.

13. Show that if $G$ is a simple graph with $n$ vertices and $k$ components, then $G$ has at most $\frac{(n-k)(n-k+1)}{2}$ edges.

14. Show that a graph is bipartite if and only if it contains no odd cycle.

## 5.4 Euler and Hamilton Paths

We introduced the Königsberg bridge problem in the beginning of this chapter, which was solved by Euler in 1736. Euler studied this problem using the multigraph obtained when the four regions were represented by vertices and the bridges by edges. The resulting graph is depicted in Figure 5.4.1.

The problem of traveling across every bridge exactly once can be represented by the following question: Does there exist a closed trail in this multigraph that contains every edge? Before answering the question, we need to introduce some definitions.

**DEFINITION 5.4.1** Suppose that $G$ is an undirected graph. A trail that traverses every edge of $G$ is called an **Euler trail**.

**DEFINITION 5.4.2** A **tour** of $G$ is a closed walk that traverses each edge of $G$ at least

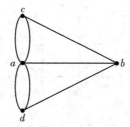

**Figure 5.4.1  The multigraph of Königsberg bridge problem**

once. An **Euler tour** is a tour which traverses each edge exactly once (so it is a closed Euler trail). A graph is an **eulerian** if it contains an Euler tour.

**EXAMPLE 5.4.1**  Which of the undirected graphs in Figure 5.4.2 is an eulerian? Of those that are not, which have an Euler trail?

*Solution*: The graph $G_1$ has an Euler tour *abdfghjigedca*. Therefore, $G_1$ is an eulerian. Neither graph $G_2$ nor graph $G_3$ has an Euler tour. However, while $G_2$ does not have any Euler trails, $G_3$ has an Euler trail, namely *cabedbcd*.

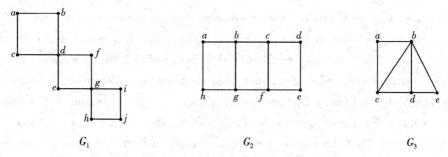

**Figure 5.4.2  Graphs for EXAMPLE 5.4.1**

**EXAMPLE 5.4.2**  Which of the directed graphs in Figure 5.4.3 are eulerian? Of those that are not, which have an Euler trail?

*Solution*: $G_1$ has an Euler trail, namely, *abcdenmkjirnjbhgef*. Neither $G_2$ nor $G_3$ has an Euler trail.

The following theorem is useful in finding Eulerians.

**THEOREM 5.4.1**  A nonempty connected graph is an eulerian if and only if it has no vertices of odd degrees.

*Proof*: Let $G$ be an eulerian, and let $C$ be an Euler tour of $G$ with origin (and

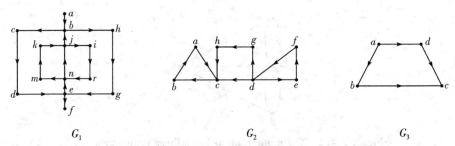

**Figure 5.4.3   Graphs for EXAMPLE 5.4.2**

terminus) $s$. For any other vertex $v$ of $G$, each time the Euler tour $C$ comes to $v$ it then departs from the vertex. Thus $C$ traverses either two new edges that are incident with $v$. Hence, if the tour passes through a vertex it contributes two to the vertex's degree. Since an Euler tour contains every edge of $G$, $d(v)$ is even for all $v \neq s$. Similarly, since $C$ starts and ends at $s$, the first edge of the tour must be distinct from the last edge, and because any other visit to $s$ results in a count of two for $d(s)$, $d(s)$ is also even. Therefore $G$ has no vertices of odd degrees.

Conversely, suppose that $G$ is a connected graph and the degree of every vertex of $G$ is even. We will construct an Euler tour. Since the minimum degree of $G$ is at least two, there exists a cycle in $G$ (by Exercise 11(b) in Section 5.3). We select a closed trail $C$. If $C$ contains every edge of $G$, then $C$ is an Euler tour. Otherwise, consider the subgraph $H$ obtained from $G$ by removing all edges of $C$. Also remove any vertex from $H$ that became isolated after the removal of edges of $C$. Then every vertex in $H$ has even degree (because in $G$ all vertices have even degree, and for each vertex removed to form $H$, pair of edges incident with this vertex have been deleted), but it may not be connected. Since $G$ is connected so $V(H) \cap V(C) \neq \emptyset$, namely there is $w \in V(H) \cap V(C)$. Firstly, we select a closed trail $C'$ in $H$ starting at $w$. Next, form a larger closed trail in $G$ by splicing the closed trail $C'$ at $w$ in $H$ with the original closed trail $C$ in $G$. Continue this process until all edges have been used. We have constructed an Euler tour.

**EXAMPLE 5.4.3**   Construct an Euler tour for the Mohammed's Scimitars graph $G$ shown in Figure 5.4.4.

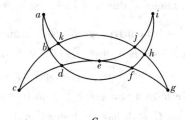

$$G$$

**Figure 5.4.4  Mohammed's Scimitars**

*Solution*: Firstly, we select a closed trail $T_1 = akedcba$. We then obtain a subgraph $H_1$ by deleting the edges in this trail and all vertices that became isolated when these edges were removed. Secondly, we form a closed trail $T_2 = efghije$ in $H_1$. Splicing the second trail $T_2$ and the first trail $T_1$ together at the common vertex $e$ results in a larger closed trail $T' = akefghijedcba$. Meanwhile, we obtain a subgraph $H_2$ of $H_1$ by deleting the edges in trail $T_2$ and all vertices that became isolated when these edges were removed. Then we have a closed trail $T_3 = bdfhjkb$. Inserting $T_3$ into $T'$ at the appropriate place produces the Euler tour $akefghijedcbdfhjkba$.

**DEFINITION 5.4.3**  Let $e$ be an edge of graph $G$. If $w(G-e) > w(G)$, then $e$ is called a **cut edge** of graph $G$.

There exists a good algorithm to determine an Euler tour in an eulerian. The algorithm, named after Fleury who constructed an Euler tour by tracing out a trail, is subject to the condition that, at any stage, a cut edge of the untraced subgraph is selected only if there is no alternative.

**Fleury's Algorithm**:
  1. Choose an arbitrary vertex $u_0$, and set $T_0 = u_0$.
  2. If the trail $T_i = u_0 e_1 u_1 \cdots e_i u_i$ has been chosen. Then select an edge $e_{i+1}$ from $E \setminus \{e_1, e_2, \cdots, e_i\}$ in such a way that
     (i) $e_{i+1}$ is incident with $u_i$;
     (ii) unless there is no alternative, $e_{i+1}$ is not a cut edge of $G_i = G - \{e_1, e_2, \cdots, e_i\}$.
  3. Stop when step 2 can no longer be implemented.

**COROLLARY 5.4.1**  If $G$ is a connected graph, then we can construct an Euler trail in $G$ if and only if $G$ has at most two vertices of odd degree.

*Proof*. If $G$ has an Euler trail then, as in the proof of THEOREM 5.4.1, each vertex other than the origin and terminus of this trail has even degree.

Conversely, suppose that $G$ is a nontrivial graph with at most two vertices having odd degree. If $G$ has no such vertices then, by THEOREM 5.4.1, $G$ has a closed Euler trail. Otherwise, $G$ has exactly two vertices, $a$ and $b$, with odd degrees. Add an additional edge $ab$ to $G$. We have a new graph $G + ab$ that is connected and has every vertex of even degree. Hence $G + ab$ has an Euler tour $C$, and when the edge $ab$ is removed from $C$, we obtain an Euler trail for $G$.

**EXAMPLE 5.4.4** Which graphs shown in Figure 5.4.5 have an Euler trail?

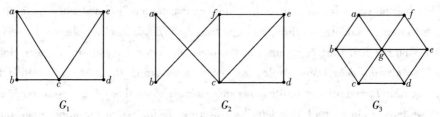

**Figure 5.4.5** Graphs for EXAMPLE 5.4.4

*Solution*: $G_1$ contains exactly two vertices of odd degree, namely, $a$ and $e$. Hence it has an Euler trail *acedcbae*. Similarly, $G_2$ has exactly two vertices of odd degree, namely, $f$ and $e$. So $G_2$ has an Euler trail that must have $f$ and $e$ as its ends. One such Euler trail is *fbacdefce*. However, $G_3$ has no Euler trail since it has six vertices of degree 3.

Returning now to the seven bridges problem of Königsberg, we know that Figure 5.4.1 is a connected graph. But it has four vertices of odd degree. Consequently, it has no Euler trail or Euler tour.

If $G$ is a digraph, then we have the following theorem.

**THEOREM 5.4.2** Let $G$ be a weakly connected and directed multigraph. $G$ has an Euler tour if and only if the in-degree and out-degree of each vertex are equal. $G$ has an Euler trail but has no Euler tour if and only if the in-degree and out-degree are equal for all but two vertices, of which one vertex has in-degree one larger than its out-degree and the other vertex has out-degree one larger than its in-degree.

The proof of this theorem is left to readers.

Euler trails and Euler tours can be used to solve many practical problems. For example, a post man picks up mails at the post office, delivers them, and then returns to the post office. He must cover each street in his area at least once. Subject to this condition, he wishes to choose his route in such a way that he walks as little as possible. This problem is known as the Chinese post man problem in honor of Guan Meigu (1962). If the graph that represents the streets the post man needs to cover is an eulerian, the Fleury's algorithm can solve the Chinese postman problem. Otherwise, if no Euler tour exists, some streets will have to be traversed more than once. Edmonds and Johnson (1973) proposed a good algorithm for this case. Unfortunately, it is too involved to be presented here. However, if the graph has exactly two vertices of odd degree, then we need to find a shortest path between these two vertices of odd degree and duplicate each edge of this shortest path. The resulting graph is an eulerian. We will introduce Dijkstra's algorithm to find a shortest path between two vertices in next section.

We have developed necessary and sufficient conditions for the existence of tours and trails that contain every edge of a multigraph exactly once. Can we do the same for paths and cycles that contain every vertex of the graph exactly once? We will now deal with this question.

**DEFINITION 5.4.4** A path that contains every vertex of $G$ is called a **Hamilton path** of $G$. A **Hamilton cycle** of $G$ is a cycle that contains every vertex of $G$. A graph is a **Hamiltonian** if it contains a Hamilton cycle.

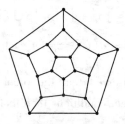

**Figure 5.4.6  The dodecahedron**

This terminology is named after Irish mathematician Sir William Roman Hamilton (1856), who invented a game on a dodecahedron (a polyhedron with 12 regular pentagons as faces shown in Figure 5.4.6). One person sticks five pins in any five consecutive vertices on a dodecahedron and another one is required to complete the graph so that it forms a spanning cycle. The dodecahedron is a Hamiltonian.

Is there a simple way to determine whether a graph has a Hamilton cycle or Hamilton path? In contrast with the case of eulerian graphs, no nontrivial necessary and sufficient conditions for a graph to be a Hamiltonian are known. However there are many known theorems that give either sufficient or necessary conditions for existence of Hamilton cycles or Hamilton paths in a connected graph. We start with a result due to Dirac (1952).

**THEOREM 5.4.3** If $G = (V, E)$ is a simple graph with $n(\geqslant 3)$ vertices and $d(v) \geqslant \frac{n}{2}$ for each vertex $v \in V$, then $G$ is a Hamiltonian.

*Proof*: We prove the theorem by contradiction. Suppose the theorem is false, and let $G$ be a maximal non-Hamiltonian simple graph with $n$ vertices and $d(v) \geqslant \frac{n}{2}$ for all $v \in V$.

Then for any edge $e$ not in $G$, $G + e$ has a Hamilton cycle. Since $n \geqslant 3$, $G$ cannot be complete. Let $u$ and $v$ be nonadjacent vertices in $G$. $G + uv$ is a Hamiltonian. Since $G$ is non-Hamiltonian, each Hamilton cycle of $G + uv$ must traverse the edge $uv$. Hence there is a Hamilton path $v_1 v_2 \cdots v_n$ in $G$ with origin $v_1$ and terminus $v_n$. Set
$$S = \{v_{i-1} \mid v_1 v_i \in E\} \text{ and } T = \{v_i \mid v_i v_n \in E\}$$
Since $v_n \notin S \cup T$, we have that $|S \cup T| < n$.

Meanwhile, for any $2 \leqslant i \leqslant n - 1$, if the edge $v_1 v_i \in E$, then $v_{i-1} v_n$ cannot be an edge of $G$, otherwise we would get a Hamilton cycle $v_1 v_2 \cdots v_{i-1} v_n v_{n-1} \cdots v_{i+1} v_i v_1$. It follows that $S \cap T \neq \emptyset$. Therefore we have
$$d(v_1) + d(v_n) = |S| + |T| = |S \cup T| - |S \cap T| < n$$
This contradicts the hypothesis that $d(v) \geqslant \frac{n}{2}$ for each vertex $v \in V$.

Bondy and Chvatal (1974) observed that the proof of THEOREM 5.4.3 can be modified to yield stronger sufficient conditions than that obtained by Dirac. Their result is described below.

**THEOREM 5.4.4** Let $G$ be a simple graph with $n$ vertices and let $u$ and $v$ be nonadjacent vertices in $G$ such that $d(u) + d(v) \geqslant n$. Then $G$ is a Hamiltonian if and only if $G + uv$ is a Hamiltonian.

*Proof*: If $G$ is a Hamiltonian, so is $G + uv$. Conversely, suppose that $G + uv$ is a Hamiltonian but $G$ is not. Then, as in the proof of THEOREM 5.4.3, we would obtain that $d(u) + d(v) < n$. But this contradicts the hypothesis.

THEOREM 5.4.4 motivates a new definition. The **closure** of $G$ is a graph obtained from $G$ by recursively joining pairs of nonadjacent vertices whose degree sum is greater than or equal to the number of vertices of $G$ until no such pair remains.

**THEOREM 5.4.5** A simple graph is a Hamiltonian if and only if the closure is a Hamiltonian.

*Proof*: Apply THEOREM 5.4.4 each time an edge is added in the formation of the closure.

THEOREM 5.4.5 has an immediate corollary.

**COROLLARY 5.4.2** Let $G = (V, E)$ be a simple graph with $|V| = n \geq 3$. If $d(u) + d(v) \geq n$ for all nonadjacent vertices $u, v \in V$, then $G$ is a Hamiltonian.

**COROLLARY 5.4.3** If $G = (V, E)$ is a simple graph with $|V| = n \geq 3$, and if $|E| = \binom{n-1}{2} + 2$, then $G$ is a Hamiltonian.

*Proof*: Let $u, v \in V$ and $uv \notin E$, and $G' = G - \{u, v\}$. Then $|E(G)| = |E(G')| + d(u) + d(v)$, because $uv \notin E$. Since $|V(G')| = n - 2$, $G'$ is a subgraph of the complete graph $K_{n-2}$. So

$$|E(G')| \leq \binom{n-2}{2}.$$

Also

$$\binom{n-1}{2} + 2 \leq |E(G)| = |E(G')| + d(u) + d(v) \leq \binom{n-2}{2} + d(u) + d(v),$$

and we find that

$$d(u) + d(v) \geq \binom{n-1}{2} + 2 - \binom{n-2}{2} = \frac{1}{2}(n-1)(n-2) + 2 - \frac{1}{2}(n-2)(n-3) = n$$

It follows from COROLLARY 5.4.2 that graph $G$ is a Hamiltonian.

In order to see the application of Hamiltonian cycle, we need to understand the concept of weighted graph. With each edge $e$ of $G$ associate a real number $W(e)$, called the weight of $e$. Then $G$, together with the weights on its edges, is called a **weighted graph**.

If $H$ is a subgraph of a weighted graph, the weight $W(H)$ of $H$ is the sum of the weights of its edges $\sum_{e \in E(H)} W(e)$.

Hamilton paths and cycles can be used to solve practical problems. For example, a traveling salesman wishes to visit a number of towns and then returns to his starting point. Given the traveling times between towns, how should he plan his traveling route so that he visits each town exactly once and travels in all for a time as short as possible? This is known as the traveling salesman problem. In graphical terms, the aim is to find a minimum weighted Hamilton cycle in a weighted complete graph.

The most straight forward way to solve an instance of the traveling salesman problem is to examine all possible Hamilton cycles and select one with minimum weight. How many cycles do we have to examine to solve the problem if there are $n$ vertices in the graph? Once a starting point is chosen, there are $(n-1)!$ different Hamilton cycles to examine, since there are $n-1$ choices for the second vertex, $n-2$ choices for the third vertex, and so on. Since a Hamilton cycle can be traveled in reverse order, we need only to examine $\frac{(n-1)!}{2}$ cycles to find our answer. However, trying to solve a traveling salesman problem in this way when there are a few dozen vertices is impractical. For example, with 25 vertices, a total of $\frac{24!}{2}(\approx 3.1 \times 10^{23})$ different Hamilton cycles would have to be considered. If it takes just one nanosecond ($10^{-9}$ second) to examine each Hamilton cycles, a total of approximately ten million years would be required to find a minimum weighted Hamilton cycle in this graph by exhaustive search technique. It is therefore desirable to have a method for obtaining a reasonable good (but not necessarily optimal) solution. One possible approach is to first find a Hamilton cycle $C$, and then search for another of smaller weight by suitably modifying $C$.

Let $C = v_1 v_2 \cdots v_n v_1$. Then for all $i$ and $j$ such that $1 < i+1 < j < n$, we can obtain a new Hamilton cycle $C_{ij} = v_1 v_2 \cdots v_i v_j v_{j-1} \cdots v_{i+1} v_{j+1} v_{j+2} \cdots v_n v_1$ by deleting the edges $v_i v_{i+1}$ and $v_j v_{j+1}$ and adding the edges $v_i v_j$ and $v_{i+1} v_{j+1}$. If, for some $i$ and $j$, the sum of weights of edges $v_i v_j$ and $v_{i+1} v_{j+1}$ is less than the sum of weighs of edges $v_i v_{i+1}$ and $v_j v_{j+1}$, then the cycle $C_{ij}$ will be an improvement of $C$.

After performing a sequence of the above modifications, we are left with a cycle that can be improved no more by these methods. This final cycle may not be optimal, but it will be fairly good. For more accuracy, the procedure can be repeated several times, starting with a different cycle each time.

As examples let us consider two applications of Euler tour and Hamilton cycle.

**EXAMPLE 5.4.5** In Figure 5.4.7(a) we have the surface of a rotating drum that is divided into sixteen sectors of equal area. In part (b) of the figure we have placed conducting (the shaded sectors and the inner circle) and non-conducting (un-shaded sectors) material on the drum. When the four terminals (shown in the figure) make contact with the four designated sectors, the non-conducting material results in no flow of current and a 0 appears on the display of a digital device. For the sectors with the conducting material, a flow of current takes place and an 1 appears on the display in each case. If the drum was rotated 22.5 degree clockwise, the screen would read 1010 (from bottom to top). So we can obtain at least two (namely 1101 and 1010) of the sixteen binary representations from 0000 to 1111. But can we represent all sixteen of them as the drum continues to rotate? And can we extend the problem to $2^n$ $n$-bit binary representations from $00 \cdots 0$ to $11 \cdots 1$?

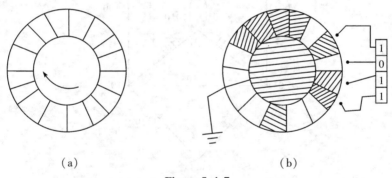

(a)            (b)

**Figure 5.4.7**

To solve the problem in the figure, we construct a directed graph $G = (V, E)$, where $V = \{000, 100, 010, 001, 011, 101, 110, 111\}$ and $E$ is constructed as follows: if $a_1 a_2 a_3, a_2 a_3 a_4 \in V$, draw an edge from $a_1 a_2 a_3$ to $a_2 a_3 a_4$. This results in the directed graph shown in Figure 5.4.8, where $|E| = 16$. We see that this graph is strongly connected and that for all $v \in V$, $d^+(v) = d^-(v) = 2$. Consequently, by THEOREM 5.4.2, it has a directed Euler tour. One Euler tour is given by

1101 1010 0101 1011 0111 1111 1110 1100 1000 0000 0001 0010 0100 1001
110→101→010→101→011→111→111→110→100→000→000→001→010→100→001

0110            011 ←            0011

The label on each edge $e = uv$, as shown in Figure 5.4.8, is a 4-bit sequence $a_1a_2a_3a_4$, where $u = a_1a_2a_3$, $v = a_2a_3a_4$. Since the vertices of $G$ are the eight distinct 3-bit sequence 000, 100, 010, 001, 011, 101, 110, 111, the labels on the sixteen edges of $G$ determine the sixteen distinct 4-bit sequences. Also, any two consecutive edge labels in the Euler tour are in the form of $a_1a_2a_3a_4$ and $a_2a_3a_4a_5$.

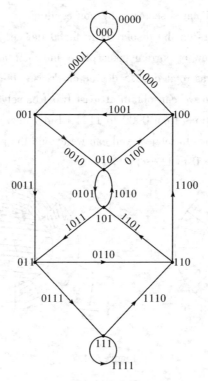

**Figure 5.4.8**

Starting with the edge 1101, in order to get next label 1010, we concatenate the last bit in 1010, namely 0, to the string 1101. The resulting string 11010 then provides 1101 (1̲1̲0̲1̲0) and 1010 (1 1̲0̲1̲0̲). The next edge label is 1010, so we concatenate the 1 (the last bit in 0101) to present string 11010 and get 110101, which provides the three distinct 4-bit sequences 1101 (1̲1̲0̲1̲01), 1010 (1 1̲0̲1̲0̲1), and 0101 (11 0̲1̲0̲1̲). Continuing in this way, we arrive at the 16-bit sequence 1101011110000100, and these sixteen bits are then arranged in the sectors of the rotating drum as in Figure 5.4.9. It is from this figure that the result in Figure 5.4.7(b) is obtained.

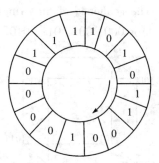

**Figure 5.4.9**

**EXAMPLE 5.4.6 (Gray code)** The position of a rotating pointer can be represented in digital form. One way to do this is to divide the circle into $2^n$ arcs of equal length and to assign a bit string of length $n$ to each arc. The digital representation of the position of the pointer can be determined using a set of $n$ contacts. Each contact is used to read one bit in the digital representation of the position. This is illustrated in Figure 5.4.10(b) for the assignment from Figure 5.4.10(a). The assignment of bit strings to the $2^n$ arcs should be made so that only one bit is different in the bit strings represented by adjacent arcs. Figure 5.4.10(a) displays such a situation in the coding scheme.

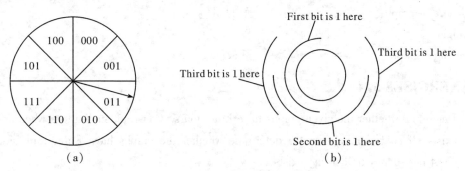

**Figure 5.4.10**

A **Gray code** is a labeling of the arcs of the circle so that adjacent arcs are labeled with bit strings that differ in exactly one bit. The assignment in Figure 5.4.10 is a Gray code. We can find a Gary code by listing all bit strings of length $n$ in such a way that each string differs in exactly one position from the preceding bit string, and the last string differs from the first in exactly one position. We can model this problem using the $n$-cube. What is needed to solve this problem is a Hamilton cycle in $n$-cube. Such Hamilton cycles are easy to find. For instance, a Hamilton cycle for a 3-cube is displayed in Figure 5.4.11. The sequence of bit strings differing in exactly one bit produced by this Hamilton cycle is 000, 001, 011, 010, 110, 111, 101, 100.

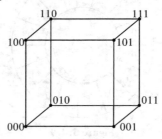

**Figure 5.4.11**

## WORDS AND EXPRESSIONS

| | |
|---|---|
| Euler trail | 欧拉迹 |
| tour | 环游 |
| eulerian | 欧拉图 |
| Hamilton cycle | 哈密尔顿圈 |
| Hamiltonian | 哈密尔顿图 |
| closure of graph | 图的闭包 |
| weight | 权 |
| weighted graph | 赋权图 |

## EXERCISES 5.4

1. Determine whether the given graphs have Euler tours. Construct such a tour when one exists. If no Euler tour exists, determine whether the graphs have Euler trails and construct such a trail if one exists.

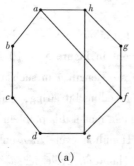

(a)

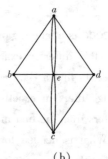

(b)

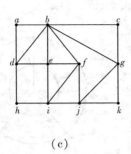
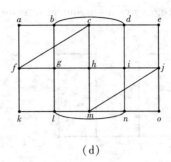

(c)                 (d)

2. Determine whether the directed graphs shown have Euler tours. Construct an Euler tour if one exists. If no Euler tour exists, determine whether the directed graphs have Euler trails. Construct an Euler trail if one exists.

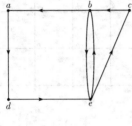
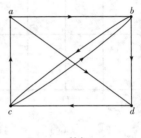

(a)                 (b)

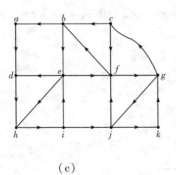
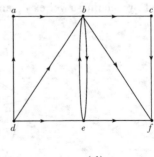

(c)                 (d)

3. Devise an algorithm for constructing Euler tours in a directed graph.

4. Show that if $G$ has $2k$ ($k>0$) vertices with odd degrees, then there are $k$ edge-disjoint tails $T_1, T_2, \cdots, T_k$ in $G$ such that $E(G) = E(T_1) \cup E(T_2) \cup \cdots \cup E(T_k)$.

5. Determine whether the given graphs below have Hamilton cycles. If yes, find such a

cycle. If no, give an argument to show why no such cycle exists.

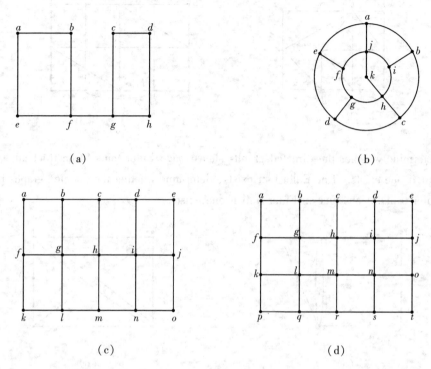

6. Show that the Petersen graph below does not have a Hamilton cycle, but the subgraph obtained by deleting a vertex $v$, and all edges incident with $v$, does have a Hamilton cycle.

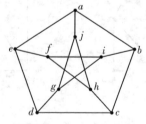

7. Let $G = (V, E)$ be a 6-regular simple graph. Prove that if $|V| = 11$, then $G$ contains a Hamilton cycle.

8. Let $G = (V, E)$ be an $n$-regular simple graph with $|V| \geq 2n + 2$. Prove that $\overline{G}$ (The complement of $G$) has a Hamilton cycle.

9. Show that there is a Gray code of order $n$ whenever $n$ is a positive integer, or equivalently, show that the $n$-cube $(n>1)$ always has a Hamilton cycle.

## 5.5 The Shortest-Path Problem

Many problems can be modeled using weighted graphs. In an airline system, for example, vertices denote the cities and edges represent flights, and weights may indicate distances between cities, or flight time, or fares.

Many optimization problems amount to finding, in a weighted graph, a subgraph of certain type with minimum (or maximum) weight. One such problem is the shortest-path problem: given an airline system, determine a shortest route between two specified cities in the system.

Here one must find, in a weighted graph, a path of minimum weight connecting two specified vertices $u_0$ and $v_0$. The weights represent distances by flight between directly linked cities, and are therefore non-negative. The path indicated by bold line in the graph of Figure 5.5.1 is an $(u_0, v_0)$-path of minimum weight.

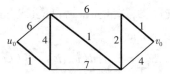

**Figure 5.5.1**

We shall refer to the weight of a path in a weighted graph as its length. The minimum weight of an $(u, v)$-path is called the distance between $u$ and $v$ and denoted by $d(u,v)$. When all weights of edges are equal to one, the definition coincides with the usual notation of length defined in Section 5.3.

We shall assume that graph $G = (V, E)$ is simple (Although the algorithm proposed is also suitable for other types of graphs). We shall also assume that all the weights are positive. We define $W(e) = 0$ if the two ends of edge $e$ are identical, and $W(uv) = \infty$, if $uv \notin E$.

We shall introduce a version of the algorithm that was discovered by Edsger Wybe Dijkstra in 1959. It finds not only a shortest path between $u_0$ and $v_0$, but shortest paths from $u_0$ to all other vertices of $G$. The basic idea is as follows.

Let $S \subset V$ and $u_0 \in S$, and denote $\overline{S} = V - S$. If $P = u_0 \cdots v_m v_{m+1}$ is a shortest path from $u_0$ to $\overline{S}$, then there exist $v_m \in S$ and $v_{m+1} \in \overline{S}$ such that the $(u_0, v_m)$-section of $P$ must be a shortest $(u_0, v_m)$-path. Therefore

$$d(u_0, v_{m+1}) = d(u_0, v_m) + W(v_m v_{m+1})$$

and the distance from $u_0$ to $\overline{S}$ is given by the formula

$$d(u_0, \overline{S}) = \min_{\substack{u \in S \\ v \in \overline{S}}} \{d(u_0, u) + W(uv)\}. \qquad (5.5.1)$$

The Dijkstra's algorithm is based on the above formula and relies on a series of iterations. A distinguished set of vertices is constructed by adding one vertex at each iteration. Starting with the set $S_0 = \{u_0\}$, an increasing sequence $S_0, S_1, \cdots, S_{n-1}$ is constructed in such a way that, at the end of stage $i$, the shortest paths from $u_0$ to all vertices in $S_i$ are known.

We begin with $S_0 = \{u_0\}$ then determine a vertex nearest to $u_0$:

$$d(u_0, \overline{S}_0) = \min_{\substack{u \in S_0 \\ v \in \overline{S}_0}} \{d(u_0, u) + W(uv)\} = \min_{v \in \overline{S}_0} \{W(u_0 v)\}.$$

If $u_1 \in \overline{S}_0$ and $d(u_0, u_1) = d(u_0, \overline{S}_0)$, then we enlarge $S_0$ to get a distinguished set $S_1 = S_0 \cup \{u_1\} = \{u_0, u_1\}$ and determine

$$d(u_0, \overline{S}_1) = \min_{\substack{u \in S_1 \\ v \in \overline{S}_1}} \{d(u_0, u) + W(uv)\}.$$

This leads us to a vertex $u_2$ in $\overline{S}_1$ with $d(u_0, \overline{S}_1) = d(u_0, u_2)$. Hence a distinguished set $S_2 = S_1 \cup \{u_2\} = \{u_0, u_1, u_2\}$ is obtained. Continuing the process, if the set $S_i = \{u_0, u_1, u_2, \cdots, u_i\}$ has been constructed, we compute $d(u_0, \overline{S}_i)$ with formula $(5.5.1)$ and select a vertex $u_{i+1} \in \overline{S}_i$ such that $d(u_0, u_{i+1}) = d(u_0, \overline{S}_i)$. The set $S_{i+1}$ is formed from $S_i$ by adding the vertex $u_{i+1}$, namely, $S_{i+1} = S_i \cup \{u_{i+1}\}$. The process stops when we reach $\overline{S}_{n-1} = \emptyset$.

In order to avoid many repeated comparisons arising from formula $(5.5.1)$, and to retain computational information, we adopt the following labeling procedure. Throughout this process, each vertex $v$ carries a label. It begins by labeling $u_0$ with 0 and the other vertices with $\infty$ (in actual computations $\infty$ is replaced by a sufficiently large number). We use the notation $L_0(u_0) = 0$ and $L_0(v) = \infty$ for these labels before any iteration has taken place (the subscript 0 stands for the "0th" iteration). The labels are the lengths of the shortest paths from $u_0$ to the vertices, while the paths contain only the vertex $u_0$ at the beginning (since no path from $u_0$ to a vertex different from $u_0$ exists, $\infty$ is the length of a shortest path from $u_0$ to such a vertex). As the algorithm proceeds, these labels are modified at each stage. At the end of stage $i$, once $u_i$ is added to $S_i$, we update two sorts of labels. The first is that for any vertex $u \in S_i$ a label $(L_i(u), x)$ is attached, where $L_i(u) = d(u_0, u)$ is the distance from $u_0$ to $u$, and $x$ is the vertex that precedes $u$ along the shortest path. Namely, $x$ satisfies $L_i(u) = d(u_0, x) + W(xu)$. The second is that for any vertex $v \in \overline{S}_i$ a label $(L_i(v), y)$ is attached where $L_i(v) = \min\{L_{i-1}(v), d(u_0, u_i) + W(u_i v)\}$, and $y$ is the vertex in $S_i$ that produces the minimum.

### Dijkstra's algorithm

1. Set $L(u_0) = 0$, $L(v) = \infty$ for $v \neq u_0$, $S_0 = \{u_0\}$ and $i = 0$.
2. For each $v \in \overline{S}_i$, $L(v) := \min\{L(v), L(u_i) + W(u_i v)\}$. Label $v$ with $(L(v), x)$ such that vertex $x$ of $S_i$ produces $L(v)$. Compute $\min_{v \in \overline{S}_i}\{L(v)\}$ and let $u_{i+1}$ denote a vertex for which this minimum is attained. Set $S_{i+1} = S_i \cup \{u_{i+1}\}$.
3. If $i = n - 1$, stop. If $i < n - 1$, $i := i + 1$ and go to step 2 (note that $n$ is the number of vertices of graph $G$).

When the algorithm terminates, the distance from $u_0$ to $v$ is given by the final value of $L(v)$. From the second term of the label of $v$ we can find the shortest path by inverse tracing. If we want to find the distance from $u_0$ to $v_0$, we can stop as soon as some $u_j$ is equal to $v_0$.

**EXAMPLE 5.5.1** Use Dijkstra's algorithm to find the length of a shortest path between the vertices $u_0$ and $v_0$ in the weighted graph shown in Figure 5.5.1.

*Solution*: The steps used by Dijkstra's algorithm to find a shortest path between $u_0$ and $v_0$ are shown in Figure 5.5.2. At each iteration of the algorithm the vertices of the set $S_i$ are circled. The algorithm terminates when $v_0$ is circled. The shortest path from $u_0$ to $v_0$ we finally obtained is $u_0 cadbv_0$ with length 9.

## WORDS AND EXPRESSIONS

the shortest-path problem      最短路径问题
distance            距离

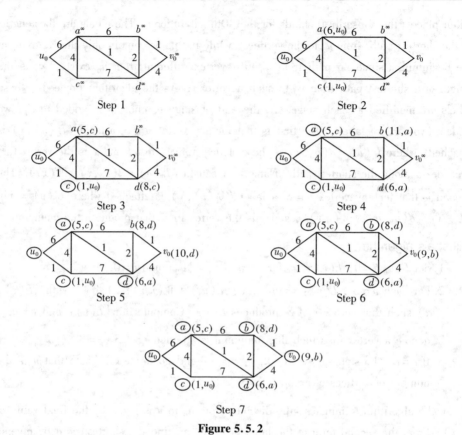

Step 7

**Figure 5.5.2**

## EXERCISES 5.5

1. Find the lengths of the shortest paths from $u_0$ to all other vertices in the given weighted graphs below.

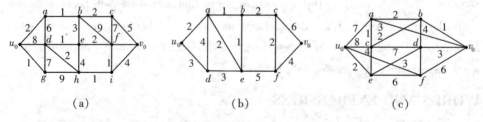

2. A company has branches in each of the five cities $C_1, C_2, C_3, C_4, C_5$. The fare for a direct flight from $C_i$ to $C_j$ is given by the $(i,j)$ th entry in the following matrix ($\infty$

indicates that there is no direct flight)

$$\begin{pmatrix} 0 & 50 & \infty & 40 & 10 \\ 50 & 0 & 15 & 20 & 25 \\ \infty & 15 & 0 & 10 & \infty \\ 40 & 20 & 10 & 0 & 25 \\ 10 & 25 & \infty & 25 & 0 \end{pmatrix}$$

The company is interested in computing a table of cheapest routes between pairs of the cities. Produce such a table.

3. A minimum total weight tour is called an optimal tour. Find an optimal tour for a postman in the following graph.

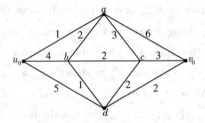

## 5.6 Planar Graphs

In this section we will discuss the question of whether a graph can be drawn in a plane without edge-crossing. For this type of graphs, we introduce the following definition.

**DEFINITION 5.6.1** A graph is called **planar** if it can be drawn in a plane with its edges intersecting only at vertices. Such a drawing is called a **planar representation** of the graph.

We sometimes refer to a planar representation of a planar graph as a plane graph.

**EXAMPLE 5.6.1** The graphs in Figure 5.6.1 are planar. The first is a 3-cube which clearly has no edge-intersecting except at the vertices. The second is $K_4$ which at the first glance appears non-planer because the edges $ad$ and $bc$ overlap at a point other than a vertex. However, $K_4$ can be drawn without crossing shown in the third graph. Therefore $K_4$ is actually planar.

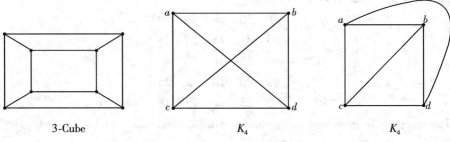

Figure 5.6.1  Planar graphs

**EXAMPLE 5.6.2**  Show that $K_5$ is non-planar.

*Proof*: It can be proved by contradiction. If $K_5$ is planar, then it is possible to find a plane graph $G$ corresponding to $K_5$. Denote the vertices of $G$ by $v_1, v_2, v_3, v_4$ and $v_5$. Since $G$ is complete, any two of its vertices are joined by an edge. Now the cycle $C = v_1 v_2 v_3 v_1$ is a closed curve that splits the plane into two regions, namely the inside of the closed curve, $R_1$, and the outside area $R_2$ as shown in Figure 5.6.2(a). The vertex $v_4$ is in either $R_1$ or $R_2$. Suppose $v_4$ is in $R_1$ (the case of $v_4 \in R_2$ can be dealt with in a similar manner), then the edges $v_1 v_4$, $v_2 v_4$ and $v_3 v_4$ divide $R_1$ into three regions $R_{11}, R_{12}$ and $R_{13}$ (see Figure 5.6.2 (b)). Now $v_5$ must live in one of the four regions $R_{11}, R_{12}, R_{13}$ and $R_2$. If $v_5 \in R_2$, then since $v_4 \in R_1$, it follows that the edge $v_4 v_5$ must meet $C$ in some point. But this contradicts the assumption that $G$ is a plane graph. The cases $v_5 \in R_{1j}, j = 1, 2, 3$ can be disposed in a like manner.

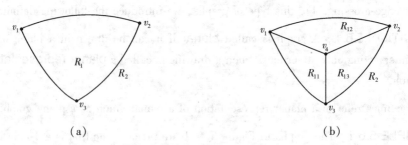

Figure 5.6.2  $K_5$ is non-planar

A similar argument can be used to establish that $K_{3,3}$, too, is non-planar. Alternatively, we shall prove that $K_5$ and $K_{3,3}$ are non-planar by another method.

A planar representation of a graph splits the plane into regions, including an unbounded region. For instance, the planar representation of $K_4$ shown in Figure 5.6.3 splits the plane into four regions: three finite areas, namely, $R_1$, $R_2$ and $R_3$, and one infinite region $R_4$. In the following theorem, Euler showed that for a given graph, all of its representations split the plane into the same number of regions.

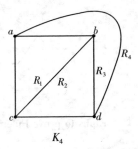

$K_4$

**Figure 5.6.3** The planar representation of $K_4$

**THEOREM 5.6.1 (Euler's Formula)** Let $G = (V, E)$ be a connected planar simple graph with $|V| = \nu$ and $|E| = \varepsilon$, $\gamma$ the number of regions in a planar representation of $G$. Then $\nu - \varepsilon + \gamma = 2$.

*Proof*: By induction on $\varepsilon$. If $\varepsilon = 1$, then $G = K_2$, and the graph has two vertices, one edge and one region (i.e., $\nu = 2$, $\varepsilon = 1$ and $\gamma = 1$). So $\nu - \varepsilon + \gamma = 2$.

Now let $k$ be a positive integer and assume that the result is true for $\varepsilon = k$. Consider any connected planar simple graph with $\nu$ vertices, $\gamma$ regions, and $\varepsilon = k + 1$ edges. Let $u, v \in V$ with $uv \in E$. Consider the subgraph $H = G - uv$. We discuss the following two cases, depending on whether $H$ is connected or disconnected.

Case 1: $H$ is connected. Then $u$ and $v$ must be on the boundary of a common region (as demonstrated in Figure 5.6.4(a)). In this situation $H$ has $\nu$ vertices, $k$ edges, and $\gamma - 1$ regions because one of the regions of $H$ has been split into two regions of $G$. The induction hypothesis applied to graph $H$ tells us that $\nu - k + (\gamma - 1) = 2$, and it follows that $2 = \nu - (k+1) + \gamma = \nu - \varepsilon + \gamma$. So Euler's Formula is true in this case.

Case 2: $H$ is disconnected. Vertices $u$ and $v$ must be in different components as demonstrated in Figure 5.6.4(b). $H$ has $\nu$ vertices, $k$ edges and $\gamma$ regions. $H$ has two components $H_1$ and $H_2$, where $H_i$ has $\nu_i$ vertices, $\varepsilon_i$ edges and $\gamma_i$ regions ($i = 1, 2$). Furthermore $\nu_1 + \nu_2 = \nu$, $\varepsilon_1 + \varepsilon_2 = k$, and $\gamma_1 + \gamma_2 = \gamma + 1$ because each of $H_1$ and $H_2$ determines an infinite region. When we apply the induction hypothesis to each of $H_1$ and $H_2$ we have that $\nu_1 - \varepsilon_1 + \gamma_1 = 2$ and $\nu_2 - \varepsilon_2 + \gamma_2 = 2$. Consequently, $(\nu_1 + \nu_2) - (\varepsilon_1$

$+\varepsilon_2) + (\gamma_1 + \gamma_2) = \nu - (\varepsilon - 1) + (\gamma + 1) = 4$, which means that $\nu - \varepsilon + \gamma = 2$, thus establishing Euler's Formula for $G$ in this case.

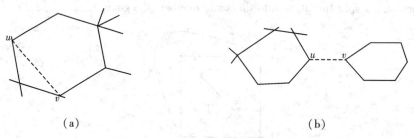

(a)　　　　　　　　　　　　(b)

**Figure 5.6.4　Graphs to prove Euler's Formula**

Let $G$ be a connected plane graph and $R$ a region of $G$. Then the boundary of $R$ can be regarded as a closed walk in which each cut edge of $G$ is traversed twice. If the boundary contains no cut edge, it is a cycle of $G$. A region $R$ is said to be incident with the vertices and edges in its boundary. If $e$ is a cut edge in a plane graph, then just one region is incident with $e$. Otherwise there are two regions incident with $e$. The degree $d(R)$ of a region $R$ is the number of edges with which it is incident (that is, the number of edges in its boundary), where cut edges are counted twice. For example, in the plane graph depicted in Figure 5.6.5, region $R_2$ has boundary $ae_1be_5de_8ee_9fe_6fe_9ee_7a$, $d(R_2) = 7$, and $R_5$ has boundary $ae_1be_2ce_3de_8ee_7a$, $d(R_5) = 5$.

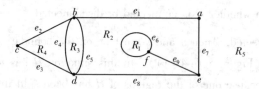

**Figure 5.6.5　A connected plane graph and its regions**

**COROLLARY 5.6.1**　If $G$ is a connected planar graph with $\varepsilon$ edges and $\nu (\geqslant 3)$ vertices, then $\varepsilon \leqslant 3\nu - 6$.

*Proof*: Let $G$ be a connected planar simple graph and $\nu \geqslant 3$. Then $d(R) \geqslant 3$ for each region $R$. Suppose that the plane graph $G$ divides the plane into $\gamma$ regions. Then
$$\sum d(R) \geqslant 3\gamma.$$
Note that $\sum d(R)$ is exactly twice the number of edges in the graph, because each

edge occurs on the boundary of a region exactly twice (either in two different regions or twice in the same region). It follows that $2\varepsilon = \sum d(R) \geq 3\gamma$. Thus, from THEOREM 5.6.1, $\nu - \varepsilon + \frac{2\varepsilon}{3} \geq 2$, which results in $\varepsilon \leq 3\nu - 6$.

**COROLLARY 5.6.2** If $G$ is a connected planar simple graph, then $G$ has a vertex whose degree is not greater than five.

*Proof*: Let $\delta$ denote the minimum degree of $G$. From handshaking theorem and COROLLARY 5.6.1, we have that $\nu\delta \leq \sum_{u \in V} d(u) = 2\varepsilon \leq 2(3\nu - 6)$, which is equivalent to $(6-\delta)\nu \geq 12$. It follows that $\delta \leq 5$.

**EXAMPLE 5.6.3** $K_5$ is non-planar. Otherwise COROLLARY 5.6.1 would give that $10 = \varepsilon(k_5) \leq 3\nu - 6 = 9$.

**EXAMPLE 5.6.4** Show that $K_{3,3}$ is non-planar.

*Proof*: Suppose that $K_{3,3}$ is a plane graph. Since $K_{3,3}$ has no cycle of length less than four, every region of $K_{3,3}$ must have a degree not less than four. Therefore, by Euler's formula and COROLLARY 5.6.1, we have that $4\gamma \leq \sum d(R) = 2\varepsilon$, $\nu - \varepsilon + \frac{\varepsilon}{2} \geq 2$, and $\varepsilon \leq 2\nu - 4$. But $\varepsilon(K_{3,3}) = 9$ and $\nu = 6$ do not satisfy the inequality $\varepsilon \leq 2\nu - 4$.

We have seen that $K_5$ and $K_{3,3}$ are not planar. Clearly, a graph is not planar if it contains any of these two graphs as a subgraph. In fact all non-planar graphs must contain a subgraph that can be obtained from $K_5$ and $K_{3,3}$ using some operations. For convenience we introduce the operation of **subdivision** of an edge. An edge $e$ is said to be **subdivided** when it is deleted and replaced by a path of length two connecting its ends. The internal vertex of this path is a new vertex. This is illustrated in Figure 5.6.6.

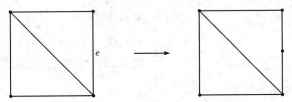

Figure 5.6.6 Subdivision of an edge

**THEOREM 5.6.2 (Kuratowski's Theorem)** A graph is planar if and only if it contains no subdivision of $K_5$ or $K_{3,3}$.

**EXAMPLE 5.6.5**  Petersen graph (Shown in Figure 5.6.7(a)) is non-planar because it contains a subdivision of $K_{3,3}$ (as demonstrated in Figure 5.6.7(b)). $H$ is a subgraph of $G$ and $H$ is also a subdivision of $K_{3,3}$.

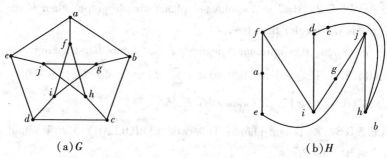

(a) $G$     (b) $H$

**Figure 5.6.7  Petersen graph and its subgraph**

Planarity of graphs plays an important role in the design of electronic circuits. We can model a circuit with a graph by representing the components of the circuit by vertices and the connections between them by edges. We can print a circuit on a single board with no connections crossing if the graph representing the circuit is planar. When the graph is not planar, we have to turn to more expensive options. One way, for example, is to partition the circuit graph into planar subgraphs, and then construct the circuit with multiple layers. We can also construct the circuit using insulated wires for the locations where connections cross. In this case, drawing the graph with the fewest possible crossing is important.

## WORDS AND EXPRESSIONS

| | |
|---|---|
| planar | 平面的 |
| planar representation | 平面表示 |
| plane graph | 平面图 |
| subdivision | 剖分,细分 |

## EXERCISES 5.6

1. Show that $K_5 - e$ is planar for any edge $e$ of $K_5$.

2. Show that $K_{3,3} - e$ is planar for any edge $e$ of $K_{3,3}$.

3. Determine whether the given graphs below are planar. If yes, draw them so that no edge crossing happens. If no, find a subdivision of $K_5$ or $K_{3,3}$.

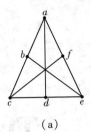

(a)

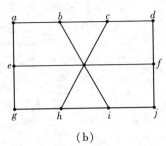

(b)

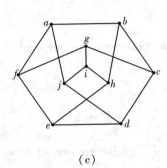

(c)

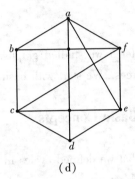

(d)

4. Suppose that a connected planar 3-regular graph has eight vertices. Into how many regions is the plane divided by a planar representation of this graph?

5. The girth of $G$ is the length of a shortest cycle in $G$. Suppose a connected planar simple graph with $\varepsilon$ edges and $\nu$ vertices has girth $k \geqslant 3$. Show that $\varepsilon \leqslant k(\nu-2)/(k-2)$.

6. Suppose that a plane graph has $w$ connected components, $\varepsilon$ edges, $\nu$ vertices, and $\gamma$ regions. Find a formula for $\gamma$ in terms of $\varepsilon, \nu$ and $w$.

# 6
# Trees

In chapter 5 we studied graphs. This chapter will focus on a special kind of graphs called trees. We start with basic definitions.

## 6.1 Basic Concepts

Recall that we defined cycles in DEFINITION 5.3.7. A basic property of a tree is that it should not contain a cycle. Let's now formally define the fundamental concepts.

**DEFINITION 6.1.1** A graph is **acyclic** if it contains no subgraph isomorphic to a cycle $C_n$.

Because acyclic graphs are very important, we give them a special name in the following definition.

**DEFINITION 6.1.2** A **forest** is an acyclic graph.

**EXAMPLE 6.1.1** Neither of the graphs shown in Figure 6.1.1 is acyclic because they contain cycles, and so neither of them is a forest.

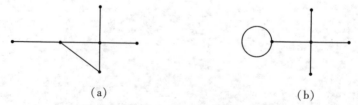

(a)          (b)

**Figure 6.1.1** Graphs containing cycles

Using the concept of cut edge defined in DEFINITION 5.4.3, it is easy to derive the following result.

**THEOREM 6.1.1**  A graph is a forest if and only if every edge in the graph is a cut edge.

Among the forests we are particularly interested in those which are connected.

**DEFINITION 6.1.3**  A **tree** is an acyclic connected graph.

**EXAMPLE 6.1.2**  The graph shown in Figure 6.1.2(a) is a forest but it is not a tree. The graph shown in Figure 6.1.2(b) is a tree.

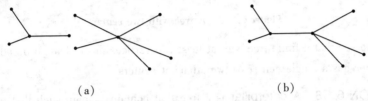

(a)                  (b)

**Figure 6.1.2  A forest and a tree**

If a tree contains only finite vertices, it is called a **finite tree**, otherwise it is infinite. In this chapter, all trees are supposed to be finite unless specified otherwise. According to DEFINITION 6.1.3 we have:

**THEOREM 6.1.2**  A connected graph with $n$ vertices is a tree if and only if it has exactly $n-1$ edges.

*Proof*: We prove the "only if" part, and leave the rest to the reader. Suppose a connected graph with $n$ vertices is a tree. Choose a vertex $r$ in the tree. We set up a one-to-one correspondence between the edges and the vertices other than $r$ by associating the terminal vertex of an edge to that edge. Since there are $n-1$ vertices other than $r$, there are $n-1$ edges in the tree.

**THEOREM 6.1.3**  A graph is a tree if and only if there is a unique simple path between any two vertices.

*Proof*: We prove the "only if" part, and leave the rest to the reader. Suppose $a$ and $b$ are two vertices of tree $T$. Since $T$ is connected, there is at least one path in $T$ that connects $a$ and $b$. If there were more than one path, then from such two paths some of the edges would form a cycle. But $T$ has no cycle.

**DEFINITION 6.1.4**  Suppose $v$ is a vertex of a graph $G$. The **eccentricity** of $v$ is the length of the longest simple path in $G$ beginning at $v$.

**DEFINITION 6.1.5** A **center** of a tree $T$ is a vertex $v$ of $T$ with minimum eccentricity. An **end vertex** (or **leaf**, or **pendent vertex**) of a tree is a vertex of degree 1.

**EXAMPLE 6.1.3** Figure 6.1.3 displays a tree with two adjacent vertices $a$ and $b$ in its center. This tree has 8 end vertices.

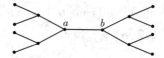

Figure 6.1.3  A tree with two centers

It is easy to see that a finite tree with at least two vertices has at least two end vertices. A tree either has a single center or two adjacent centers.

**DEFINITION 6.1.6** A **caterpillar** is a tree that contains a path such that every edge has one or two endpoints in that path.

**EXAMPLE 6.1.4** Figure 6.1.4 shows a caterpillar.

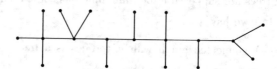

Figure 6.1.4  A caterpillar

## WORDS AND EXPRESSIONS

| | |
|---|---|
| acyclic graph | 无圈图 |
| forest | 森林 |
| eccentricity | 离心率 |

## EXERCISES 6.1

1. Prove the "if" part of THEOREM 6.1.2.
2. Prove the "if" part of THEOREM 6.1.3.

## 6.2 Roots and Orderings

Trees with special structures have vast applications in computer science. Different structures can be generated by different orderings. In this section we study some of the most useful orderings that based on rooted trees.

**DEFINITION 6.2.1** A **rooted tree** $(T, r)$ is a tree $T$ with a distinguished vertex $r$ (the **root**), in which all edges are implicitly directed away from the root. Two rooted trees $(T_1, r_1)$ and $(T_2, r_2)$ are **isomorphic as rooted trees** if there is an isomorphism $f: T_1 \to T_2$ such that $f(r_1) = r_2$.

A general tree has no directions on the edges. But according to DEFINITION 6.2.1, a rooted tree is a directed graph with the direction on each edge away from the root.

**EXAMPLE 6.2.1** The graph in Figure 6.2.1 is an example of rooted trees. Note that we use a circle to represent the root $r$.

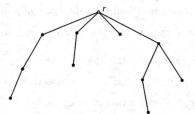

Figure 6.2.1  A rooted tree

**EXAMPLE 6.2.2** The two rooted trees in Figure 6.2.2(a) and (b) are isomorphic as graphs, but they are considered to be different as rooted trees, because there is no graph isomorphism from one to the other that maps root to root. Nevertheless the two rooted trees in Figure 6.2.2(c) and (d) are isomorphic as rooted trees.

(c)                                    (d)

Figure 6.2.2

In fact the trees in Figure 6.2.2(a) and (b) can be obtained from the same graph by choosing different vertices as roots.

**DEFINITION 6.2.2** Suppose $v$ is a vertex of a rooted tree $T$. A **child** of $v$ in $T$ is a vertex that is the immediate successor of $v$ on a path from the root. A **descendant** of $v$ in $T$ is $v$ itself or any vertex that is a successor of $v$ on a path from the root. A **proper descendant** of $v$ in $T$ is any descendant of $v$ except $v$ itself. The **parent** of $v$ in $T$ is a vertex that is the immediate predecessor of $v$ on a path to $v$ from the root. An **ancestor** of $v$ in $T$ is $v$ itself or any vertex that is the predecessor of $v$ on a path to $v$ from the root. A **proper ancestor** of $v$ in $T$ is any ancestor of $v$ except $v$ itself.

**EXAMPLE 6.2.3** In Figure 6.2.3, the rooted tree has vertices $a,b,c,d,e,f,g,h,i,j$. Their parents are $\emptyset,a,a,a,b,c,d,d,e,g$, respectively. $a$ has 3 children $b,c,d$. $d$ has 3 proper descendant $g,h,j$ and a proper ancestor $a$.

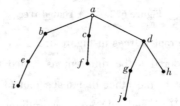

Figure 6.2.3

**DEFINITION 6.2.3** The **parent function** of a rooted tree $T$ maps the root of $T$ to the empty set and maps every other vertex to its parent.

Obviously a rooted tree can be uniquely represented by its vertex set together with its parent function, because all edges can be determined by the parent function.

**DEFINITION 6.2.4** The **siblings** in a rooted tree are those vertices with the same parent. An **internal vertex** in a rooted tree is a vertex that has children. A **leaf** in a rooted tree is a vertex that has no children. The **depth** of a vertex in a rooted tree is the number of edges in the unique path from the root to that vertex. The $n$**th level** in a rooted tree is the set of all vertices at depth $n$. The **height** of a rooted tree is the maximum depth of the vertices of the tree.

**EXAMPLE 6.2.4** The tree in Figure 6.2.4 is rooted at $a$. Vertices $e$ and $f$ are siblings. But $d$ is not a sibling of $e$ or $f$. The leaves are the vertices $d$, $f$ and $g$. The internal vertices are $a, b, c$ and $e$.

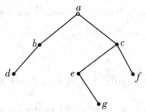

**Figure 6.2.4**

Among rooted trees there is a special kind of trees which have certain order between their vertices.

**DEFINITION 6.2.5** An **ordered tree** is a rooted tree in which the children of each internal vertex are linearly ordered.

Unless otherwise specified, a tree is generally considered as an ordered tree.

**DEFINITION 6.2.6** A **left sibling** of a vertex $v$ in an ordered tree is a sibling of $v$ that precedes $v$ in the ordering of $v$ and its siblings. A **right sibling** of a vertex $v$ in an ordered tree is a sibling of $v$ that follows $v$ in the ordering of $v$ and its siblings.

**DEFINITION 6.2.7** A **plane tree** is a drawing of an ordered tree such that the left-to-right order of the children of each vertex in the drawing is consistent with the linear ordering of the corresponding vertices in the tree.

**EXAMPLE 6.2.5** The two ordered trees in Figure 6.2.2(c) and (d) are plane trees.

**DEFINITION 6.2.8** In the **level ordering** of the vertices of an ordered tree, $u$ precedes $v$ under any of these circumstances:
1. The depth of $u$ is less than the depth of $v$;
2. $u$ is a left sibling of $v$;

3. The parent of $u$ precedes the parent of $v$.

**DEFINITION 6.2.9**   Two ordered trees $(T_1, r_1)$ and $(T_2, r_2)$ are **isomorphic** as ordered trees if there is a rooted tree isomorphism $f: T_1 \to T_2$ that preserves the ordering at every vertex.

**EXAMPLE 6.2.6**   The two ordered trees in Figure 6.2.2(c) and (d) are not isomorphic as ordered trees although they are isomorphic as rooted trees, because there is no rooted tree isomorphism from one to the other that preserves the child ordering at every vertex.

**DEFINITION 6.2.10**   An **m-ary tree** is a rooted tree such that every internal vertex has at most $m$ children. A full **m-ary tree** is a rooted tree such that every internal vertex has exactly $m$ children.

**EXAMPLE 6.2.7**   The tree in Figure 6.2.4 is a 2-ary tree of height 3.

**THEOREM 6.2.1**   Let $T$ be a tree. Let $n$, $k$, and $i$ be the number of vertices, internal vertices, and leaves of $T$.
1. If $T$ is a full $m$-ary tree then $n = mk + 1$ and $i = (m-1)k + 1$;
2. If $T$ is an $m$-ary tree of height $h$ then there are at most $m^h$ leaves in $T$.

*Proof*:
1. It is obvious that $n = k + i$. Every vertex, except the root, is the child of an internal vertex. Since each of the $k$ internal vertices has $m$ children, there are $mk$ vertices in the tree other than the root. Consequently the tree contains $mk + 1$ vertices. Therefore $n = k + i = mk + 1$, which results in $i = (m-1)k + 1$.
2. We use mathematical induction on the height. If an $m$-ary tree has height 1, then it consists of a root and no more than $m$ children, each of which is a leaf. Hence there are no more than $m^1 = m$ leaves. Assume that the result is true for all $m$-ary trees of height less than $h$. Let $T$ be an $m$-ary tree of height $h$. The leaves of $T$ are also the leaves of the subtrees of $T$ obtained by deleting the edges from the root to each of the vertices at level 1. Each of the subtrees has height less than or equal to $h - 1$. By the inductive hypothesis, each of these rooted trees has at most $m^{h-1}$ leaves. Since there are at most $m$ such subtrees, there are at most $m \cdot m^{h-1} = m^h$ leaves in $T$.

**DEFINITION 6.2.11**   A **pure binary tree** is a rooted tree such that every internal vertex has at most two children.

Normally a pure binary tree is simply called a **binary tree**. A binary tree is a 2-ary tree such that every child is distinguished as left child or right child. Binary trees are very important tools in computer science and they are also used in permutation groups vastly. For example, binary search tree is a special kind of binary tree used to implement a random access table with $O(n)$ maintenance and retrieval algorithms.

**DEFINITION 6.2.12** The **principal subtree** at a vertex $v$ of a rooted tree comprises all descendants of $v$ and all edges incident to these descendants. It has $v$ designated as its root. The **left subtree** of a vertex $v$ in a binary tree is the principal subtree at the left child. The **right subtree** of $v$ is the principal subtree at the right child.

**DEFINITION 6.2.13** A **balanced tree** of height $h$ is a rooted $m$-ary tree in which all leaves are of depth $h$ or $h-1$. A **complete binary tree** is a binary tree in which every parent has two children and all leaves are at the same depth. A **complete $m$-ary tree** is an $m$-ary tree in which every parent has $m$ children and all leaves are at the same depth.

In computer science, balanced binary trees can be used to implement a priority queue with $O(n)$ enqueue and dequeue algorithms.

**EXAMPLE 6.2.8** The tree in Figure 6.2.1 is not a balanced tree, but the tree in Figure 6.2.3 is a balanced tree of height 3.

**EXAMPLE 6.2.9** Figure 6.2.5(a) is a complete binary tree of height 2. Figure 6.2.5(b) is a complete 3-ary tree of height 2. Note that a 3-ary tree is also called a **ternary tree**.

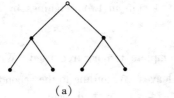
(a)
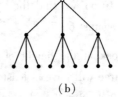
(b)

**Figure 6.2.5**

**DEFINITION 6.2.14** A **decision tree** is a rooted tree in which every internal vertex represents a decision and each path from the root to a leaf represents a cumulative choice. Decision trees can be used to solve a wide range of problems including sorting problems. The following example illustrates an application of decision trees to searching.

**EXAMPLE 6.2.10**  There are eight coins identical in appearance, and a pan balance. If exactly one of these coins is counterfeit and heavier than the other seven, how to find it with the pan balance?

*Solution*: Label the coins with 1, 2, ⋯, 8. In using the pan balance to compare sets of coins there are three outcomes to consider: (1) The two sides balance, indicating that the coins in the pans are not counterfeit; (2) The left pan goes down, indicating that the counterfeit coin is in the left pan; (3) The right pan goes down, indicating that the counterfeit coin is in the right pan.

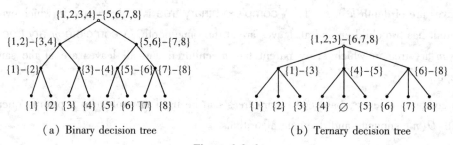

(a) Binary decision tree       (b) Ternary decision tree

**Figure 6.2.6**

In Figure 6.2.6(a) we search for the counterfeit by first balancing coins 1, 2, 3, 4 against 5, 6, 7, 8. If the balance tips to the right, we follow the right branch from the root to balance 5 and 6 against 7 and 8, and so on. At level 3 the counterfeit coin has been identified.

In Figure 6.2.6(b) we find the counterfeit coin in two weighings by first balancing coins 1, 2, 3 against 6, 7, 8.

Let's have a look at the height of the complete ternary tree used. With eight coins involved, the tree will have at least eight leaves. According to the second conclusion of THEOREM 6.2.1, $8 \leqslant 3^h$, which leads to $h \geqslant \lceil \log_3 8 \rceil = 2$. This means that at least 2 weighings are needed to find the counterfeit.

Coding theory plays an important part in many different application areas of discrete mathematics such as computer science. It enables us to represent and transmit information in terms of symbols in a given alphabet. For instance, we often represent characters internally in a computer by means of strings of the bits 0 and 1. If we want to encode the 26 letters of English alphabet, we need 5 bits since $2^4 < 26 < 2^5$. However, in English language, not all letters occur with the same frequency. It would be more efficient to use binary sequences of different lengths such that most frequently occurring letters are represented by the shortest sequences. But when letters are encoded using varying numbers of bits, some methods must be used to determine where the bits for each character start and end. For example, if *e*, *a*, and *r* were encoded with 0, 1, and 01 respectively, then the bit string 0101 could represent *ear*, *rea*, *eaea*, or *rr*.

To ensure that no bit string represents more than one sequence of letters, the bit string for a letter must never occur as the first part of the bit string for another letter. Codes with this property are called prefix codes. Formally we have the following definition.

**DEFINITION 6.2.15** A **prefix code** for a finite set $X = \{x_1, \cdots, x_n\}$ is a set $\{c_1, \cdots, c_n\}$ of binary strings in $X$ (called **codewords**) such that no codeword is a prefix of any other codeword.

For example, the encoding of *e*, *a*, and *r* as 0, 10, 11 is a prefix code. With this encoding, a word can be recovered from the unique bit string that encodes its letters. For instance, the string 10110 is the encoding of *are*.

A prefix code can be represented with a binary tree, where the characters are the labels of the leaves in the tree. The edges of the tree are labeled so that an edge leading to a left child is assigned a 0, and an edge leading to a right child is assigned a 1. The bit string used to encode a character is the sequence of labels of the edges in the unique path from the root to the leaf that has this character as its label.

**EXAMPLE 6.2.11** A binary tree with a prefix code is shown in Figure 6.2.7. Find the encoding of all the letters shown in the figure.
*Solution*: *a*: 0, *e*: 10, *i*: 1100, *o*: 1101, *u*: 111.
The tree representing a code can be used to decode a bit string. For example, the word encoded by 11111011100 using the code in Figure 6.2.7 can be uniquely decoded as *uoi*.

Obviously different encoding may have different efficiencies. There are algorithms that can be used to produce efficient codes based on the frequencies of occurrences of characters. One of the algorithms is Huffman coding.

**DEFINITION 6.2.16** A **Huffman code** for a set $X$ with a probability measure Pr is a

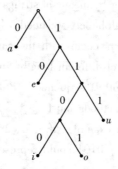

**Figure 6.2.7  A binary tree representing a prefix code**

prefix code $\{c_1, \cdots, c_n\}$ such that $\sum_{j=1}^{n} len(c_j) \Pr(x_j)$ is minimum among all prefix codes, where $len(c_j)$ measures the length of $c_j$ in bits.

Given the probabilities $\Pr(x_1), \cdots, \Pr(x_n)$ on a set $X$, we can use the following algorithm to find a Huffman code for $(X, \Pr)$.

**ALGORITHM 6.2.1**  Find a Huffman code.
1. Initialize $F$ to be a forest of isolated vertices, labeled as $x_1, \cdots, x_n$, each to be regarded as a rooted tree;
2. Assign weight $\Pr(x_j)$ to the rooted tree $x_j$ for $j = 1, \cdots, n$;
3. Repeat until forest $F$ is a single tree:
   a) choose two rooted trees, $T$ and $T'$, of smallest weights in forest $F$;
   b) replace trees $T$ and $T'$ in forest $F$ by a tree with a new root whose left subtree is $T$ and right subtree is $T'$;
   c) label the new edge to $T$ with 0 and the new edge to $T'$ with 1;
   d) assign weight $W(T) + W(T')$ to the new tree.
4. return tree $F$. The Huffman code word for $x_i$ is the concatenation of the labels on the unique path from the root to $x_i$.

**DEFINITION 6.2.17**  A **Huffman tree** for a set $X$ with a probability measure $\Pr$ is a tree constructed by Huffman's algorithm to produce a Huffman code for $(X, \Pr)$.

**EXAMPLE 6.2.12**  Given the set $X = \{a, b, c, d, e, f\}$ with respective probabilities $\{0.08, 0.10, 0.12, 0.15, 0.20, 0.35\}$, construct a Huffman tree.

*Solution*: According to ALGORITHM 6.2.1 we construct the following forests:

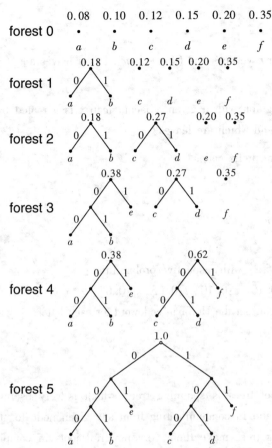

Since forest 5 above is a single tree, it is the Huffman tree we are looking for. From the tree, the Huffman code for $X = \{a, b, c, d, e, f\}$ with the given respective probabilities are 000, 001, 100, 101, 01, and 11 respectively. The most frequently used objects in the set are represented by the shortest binary codes.

## WORDS AND EXPRESSIONS

| | |
|---|---|
| rooted tree | 有根树 |
| ordered tree | 有序树 |
| binary tree | 二叉树 |
| decision tree | 决策树 |
| prefix code | 前缀码 |
| sibling | 同胞 |

# EXERCISES 6.2

1. Prove that a full $m$-ary tree with $n$ vertices has $((m-1)n+1)/m$ leaves and $(n-1)/m$ internal vertices.

2. How to determine whether a digraph with a given adjacent matrix is a rooted tree? How to know which is the root and which are leaves?

3. Which of these sets of strings are prefix codes?
   a. $A_1 = \{1,10,110,1110\}$
   b. $A_2 = \{1,01,001,000\}$
   c. $A_3 = \{b,c,aa,ac,aba,abb,abc\}$
   d. $A_4 = \{0,10,110,1000\}$

4. Given the set $X = \{a,b,c,d,e,f,g\}$ with respective probabilities:
   $a: 0.35$, $b: 0.20$, $c: 0.15$, $d: 0.10$, $e: 0.10$, $f: 0.05$, $g: 0.05$
   construct a Huffman tree, and point out the Huffman codeword for each letter.

## 6.3  Spanning Trees

One of the most important kind of trees is spanning tree, which is very useful in searching the vertices of a graph and in communicating from any given node to other nodes. These features of spanning trees make them very powerful tools in computer science. There are a number of significant results regarding this topic. But limited by the size of the book, this section will only cover the basic concepts.

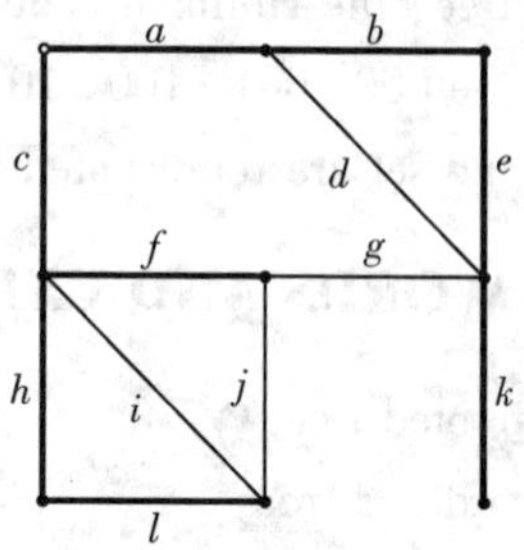

**DEFINITION 6.3.1**  Suppose $G$ is a graph. A **spanning tree** of $G$ is a tree that is a subgraph of $G$ containing every vertex of $G$. A **tree edge** of $G$ with a spanning tree $T$ is an edge $e$ such that $e \in T$. A **chord** of $G$ with a spanning tree $T$ is an edge $e$ such that $e \notin T$.

**Figure 6.3.1**

**EXAMPLE 6.3.1**  Figure 6.3.1 displays a graph and its spanning tree, in which the tree edges are $a,b,c,e,f,h,k,l$ and the chords are $d,g,i,j$.

**EXAMPLE 6.3.2** **Multicasting** plays an important role in Internet Protocol(IP) networks. There are many ways to send data from a source computer to multiple receiving computers. One way is to send the data separately to each receiver. This method, called **unicasting**, is inefficient since many copies of the data have to be transmitted over the network. With IP multicasting, a computer sends a single copy of data over the network which, upon reaching intermediate routers, is forwarded to one or more other routers so that ultimately all receivers receive the data. In multicasting, for data to reach receivers as quickly as possible, there should be no loops in the path that data take through the network. That is, once data have reached a particular router, they should never return this router again. To avoid loops, network algorithms are needed to construct a spanning tree in the graph that includes the multicast source, the routers, and the subnetworks containing receiving computers as vertices, where the edges in the graph represent the links between the computers and/or routers. In such a spanning tree, the root is the multicast source, and the subnetworks containing receiving computers are the leaves. Figure 6.3.2 shows an example.

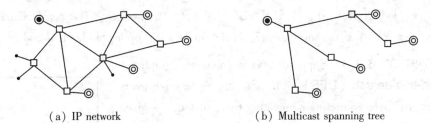

(a) IP network  (b) Multicast spanning tree

**Figure 6.3.2**  □Routers, ●Source, • Subnetworks, ◎Subnetworks with receivers

**THEOREM 6.3.1** Every connected finite graph has at least one spanning tree.

*Proof*: Suppose $G$ is a connected graph. If $G$ is not a tree, it must contain a simple cycle. Remove an edge from the cycle. The resulting subgraph has 1 fewer edge but still contains all the vertices of $G$ and remains connected. If this subgraph is not a tree, it still has a simple cycle. Remove an edge from the cycle as before. Repeat this process until no simple cycle remains. This is possible because there are only finite number of edges in $G$. The final result is a tree since the graph stays connected as edges are removed. This tree is a spanning tree of $G$ since it contains every vertex of $G$.

It is clear that there are only finitely many spanning trees in a finite graph. The following theorem, presented without proof, gives a sufficient and necessary condition regarding the number of spanning trees in a graph.

**THEOREM 6.3.2**  A connected graph $G(V,E)$ has $k$ edge-disjoint spanning trees if and only if for every partition of $V$ into $m$ nonempty subsets, there are at least $k(m-1)$ edges connecting vertices in different subsets.

For digraphs defined in DEFINITION 5.1.4, we have:

**DEFINITION 6.3.2**  Suppose $G$ is a digraph, and $T$ is a spanning tree of $G$. A **back edge** of $G$ with $T$ is a chord $e$ that joins one of its endpoints to an ancestor in $T$. A **forward edge** of $G$ with $T$ is a chord $e$ that joins one of its endpoints to a descendent in $T$. A **cross edge** of $G$ with $T$ is a chord $e$ that is neither a back edge nor a forward edge. The **fundamental cycle** of a chord $e$ with respect to $T$ consists of the edge $e$ and the unique path in $T$ joining the endpoints of $e$.

**EXAMPLE 6.3.3**  The graph shown in Figure 6.3.3 is a digraph on the same vertex and edge set of the graph of Figure 6.3.1. Chord $d$ is a forward edge, chord $i$ is a back edge, and chords $g$ and $j$ are cross edges. The fundamental cycles of the chords $d, g, i$ and $j$ are $\{d,b,e\}$, $\{g,f,c,a,b,e\}$, $\{i,h,l\}$ and $\{j,f,h,l\}$, respectively. The non-fundamental cycle $\{a,d,g,c,f\}$ is the sum (mod 2) of the fundamental cycles of chords $d$ and $g$.

**DEFINITION 6.3.3**  Suppose $G$ is a connected graph. A **depth-first search (DFS)** of $G$ is a way to traverse every vertex of $G$ by constructing a spanning tree, rooted at a given vertex $r$. Each stage of the DFS traversal seeks to move to an unvisited neighbor of the most recently visited vertex, and backtracks only if there is none available. A **depth-first-search tree (DFS-tree)** is the spanning tree constructed during a depth-first search. **Backtracking** during a depth-first search means retreating from a vertex with no unvisited neighbors back to its parent in the DFS-tree.

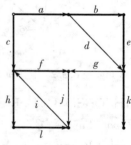

Figure 6.3.3

DFS-trees can be used to find the components, cut-points, blocks, and cut-edges of a graph. We now give an algorithm to construct the DFS-tree for a given graph.

**ALGORITHM 6.3.1**  Constructing a depth-first search spanning tree

$G$ is a connected $n$-vertex graph, and $r$ is a starting vertex.

Initialize all vertices as unvisited and all edges as unused

$E_T := \emptyset$; $loc := 1$
dfs($r$)
**procedure** dfs($u$)
    mark $u$ as visited
    $X[loc] := u$
    $loc := loc + 1$
    while vertex $u$ has any unused edges
        $e :=$ next unused edge at $u$
        mark $e$ as used
        $w :=$ the other endpoint of edge $e$
        if $w$ is unvisited then
            add $e$ to $E_T$
            dfs($w$)

The final result: $E_T$ is the edge set of the spanning tree, and array $X[1 \cdots n]$ is the list of $V(G)$ in DFS-order.

**EXAMPLE 6.3.4** Figure 6.3.4 is a re-draw of the graph of Figure 6.3.1. Suppose the local order of adjacencies at each vertex is the alphabetic order of the edge labels. If the starting vertex is the one labeled as 1, then the construction of the DFS-tree is shown in Figure 6.3.5.

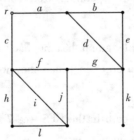

**Figure 6.3.4**

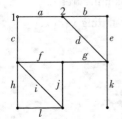 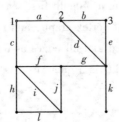 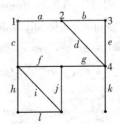

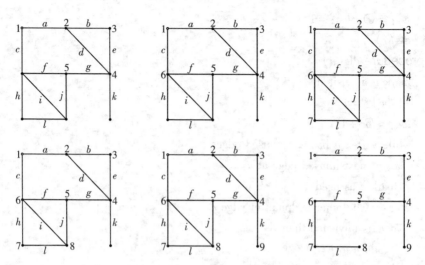

Figure 6.3.5  Constructing a depth-first search spanning tree

**DEFINITION 6.3.4**  Suppose that $G$ is a connected graph. A **breadth-first search** (**BFS**) of a graph $G$ is a way to traverse every vertex of $G$ by constructing a spanning tree, rooted at a given vertex $r$. After the BFS traversal visits a vertex $v$, all of the previously unvisited neighbors of $v$ are enqueued, and then the traversal removes from the queue the vertex at the first position of the queue, and visits that vertex. A **breadth-first-search tree** is the spanning tree constructed during a breadth-first search.

Note that a BFS-tree in a simple graph has no back edges. The level order of the vertices of an ordered tree is the order in which they would be traversed in a breadth-first search of the tree.

**THEOREM 6.3.3**  The unique path in the BFS-tree $T$ of a graph $G$ from its root $r$ to a vertex $v$ is a shortest path in $G$ from $r$ to $v$.

**ALGORITHM 6.3.2**  Constructing a breadth-first search spanning tree
$G$ is a connected locally ordered $n$-vertex graph, and $r$ is a starting vertex.
   Initialize all vertices as unvisited and all edges as unused
      $E_T := \varnothing$; $loc := 1$; $Q := r$ {$Q$ is a queue}
   while $Q \neq \varnothing$
      $x :=$ the front vertex in $Q$

    remove $x$ from $Q$
  bfs($r$)
**procedure** bfs($u$)
  mark $u$ as visited
  $X[loc] := u$
  $loc := loc + 1$
  while vertex $u$ has any unused edges
    $e :=$ next unused edge at $u$
    mark $e$ as used
    $w :=$ the other endpoint of edge $e$
    if $w$ is unvisited then
      add $e$ to $E_T$
      add $w$ to the end of $Q$

The final result: $E_T$ is the edge set of the spanning tree, and array $X[1\cdots n]$ is the list of $V(G)$ in BFS-order.

**EXAMPLE 6.3.5** Suppose in the graph of Figure 6.3.4, the local order of adjacencies at each vertex is the alphabetic order of the edge labels. Then the construction of the BFS-tree is shown in Figure 6.3.6.

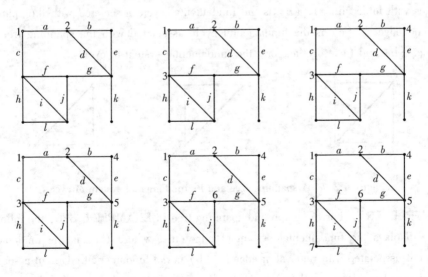

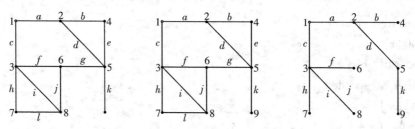

**Figure 6.3.6  BFS-tree construction**

**DEFINITION 6.3.5**  Suppose that $G$ is a connected graph, and $T$ is a spanning tree of $G$. The **fundamental cycle** of $G$ associated with $T$ and an edge $e$ of $G$ not in $T$ is the unique cycle created by adding the edge $e$ to $T$. The **fundamental system of cycles** of $G$ associated with $T$ is the set of fundamental cycles corresponding to the edges of $G$-$T$. Given two vertex sets $X_1$ and $X_2$ that partition the vertex set of $G$, the **partition-cut** $< X_1, X_2 >$ is the set of edges of $G$ that have one endpoint in $X_1$ and the other in $X_2$. The **fundamental edge-cut** of $G$ associated with removal of an edge $e$ from $T$ is the partition-cut $< X_1, X_2 >$ where $X_1$ and $X_2$ are the vertex-sets of the two components of $T-e$. The **fundamental system of edge-cuts** of $G$ associated with $T$ is the set of fundamental edge-cuts that result from removal of an edge from the tree $T$.

**EXAMPLE 6.3.6**  In Figure 6.3.7, (a) is a connected graph and its spanning tree (edges with folded lines). (b) is the fundamental cycle associated with the spanning tree and edge $d$. (c) is the fundamental cycle associated with the spanning tree and edge $e$. (b) and (c) together form the fundamental system of cycles.

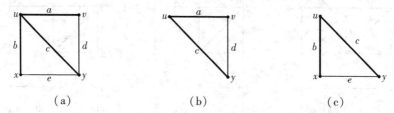

(a)                (b)                (c)

**Figure 6.3.7  A spanning tree and its fundamental system of cycles**

**EXAMPLE 6.3.7**  For the graph and spanning tree in EXAMPLE 6.3.6, the following figure displays the fundamental system of edge-cuts, where (a) is the fundamental edge-cut associated with removal of edge $a$, (b) is the fundamental edge-cut associated with removal of edge $b$, and (c) is the fundamental edge-cut associated with removal of edge $c$.

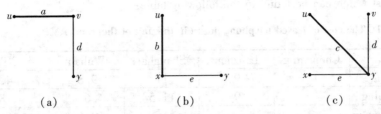

(a)          (b)          (c)

**Figure 6.3.8**   The fundamental system of edge-cuts of the spanning tree in Figure 6.3.7

It can be shown that the fundamental cycle of an edge $e$ with respect to a spanning tree $T$ such that $e \notin T$ consists of edge $e$ and those edges of $T$ whose fundamental edge-cuts contain $e$. Also, the fundamental edge-cut with respect to removal of edge $e$ from a spanning tree $T$ consists of edge $e$ and those edges of $E(G) - E(T)$ whose fundamental cycles contain $e$, where $E(G)$ and $E(T)$ denote the sets of edges of $G$ and $T$, respectively.

Based on weighted graphs introduced in Section 5.4, we have the following definition.

**DEFINITION 6.3.6**   A **minimum spanning tree** in a connected weighted graph is a spanning tree that has the smallest sum of weights of its edges.

The following algorithm can be used to find minimum spanning trees in a weighted graph.

**ALGORITHM 6.3.3**   Prim's Algorithm

$G$ is a weighted connected undirected graph with $n$ vertices.

    $T := $ a minimum-weigh edge

    For $i := 1$ to $n - 2$

    begin

       $e := $ an edge of minimum weight incident to a vertex in $T$

           and not forming a simple circuit in $T$ if add to $T$

       $T := T$ with $e$ added

    end

$T$ is a minimum spanning tree of $G$.

Prim's Algorithm can also be used to find a spanning tree for an unweighted graph by assigning a weight of 1 to each edge of the graph.

**EXAMPLE 6.3.8**   A company plans to build a communication network connecting its servers in 6 cities. Any pair of these servers can be linked with a leased telephone line with a cost which can be found in the following table:

Table 6.3.1  The cost of leased telephone lines (in the unit of thousand RMB)

|  | Chongqing | Lanzhou | Shanghai | Wuhan | Xi'an |
|---|---|---|---|---|---|
| Beijing | 14 | 9 | 11 | 9.8 | 4.5 |
| Chongqing |  | 9 | 11.5 | 6 | 8 |
| Lanzhou |  |  | 15 | 9.5 | 5.5 |
| Shanghai |  |  |  | 7 | 12 |
| Wuhan |  |  |  |  | 7.5 |

Decide which links should be made to ensure that there is a path between any two servers so that the total cost of the network is minimized.

*Solution*: The network can be represented with a weighted graph shown in Figure 6.3.9, where the weights are the costs in the unit of thousand RMB. This problem can be solved by finding a minimum spanning tree in the graph of Figure 6.3.9. The spanning tree is obtained by ALGORITHM 6.3.3 and the steps are listed below.

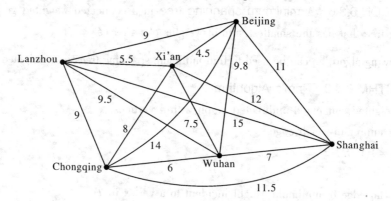

Figure 6.3.9  A weighted graph showing lease costs for a company network

| Choice | Edge | Cost |
|---|---|---|
| 1 | {Beijing, Xi'an} | 4.5 |
| 2 | {Xi'an, Lanzhou} | 5.5 |
| 3 | {Xi'an, Wuhan} | 7.5 |
| 4 | {Wuhan, Chongqing} | 6 |
| 5 | {Wuhan, Shanghai} | 7 |

The minimum spanning tree obtained is displayed in Figure 6.3.10.

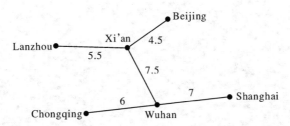

**Figure 6.3.10** The minimum spanning tree of the graph in Figure 6.3.9

Remark: Using the concept of trees, it is easy to see that Dijkstra's algorithm introduced in Section 5.5 constructs a spanning tree $T$ in an edge-weighted graph so that for each vertex $v$, the unique path in $T$ from a specified root $r$ to $v$ is a minimum-cost path from $r$ to $v$ in the graph. When all edges have unit weights, Dijkstra's algorithm produces the BFS-tree.

## WORDS AND EXPRESSIONS

| | |
|---|---|
| depth-first-search | 深度优先搜索 |
| breadth-first-search | 宽度优先搜索 |
| chord | 弦 |
| minimum spanning tree | 最小生成树 |

## EXERCISES 6.3

1. How many different spanning trees are there in the complete graph $K_5$?

2. Use Prim's algorithm to find a minimum spanning tree of the weighted graph below.

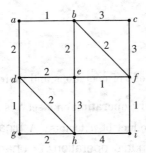

# 7

# Algebra Structures

Algebra is a branch of mathematics dealing with Arithmetic operations using variables. We have learnt the theory of logical propositions and the theory of sets in previous chapters. Both theories involve operations applied on members of certain sets. Generally speaking, some operations following certain rules, together with the sets of objects, are the subjects of algebra. For example, we formally have the names of "the algebra of sets" and "the algebra of logical propositions". A natural generalization of these two algebra is Boolean algebra which forms an abstract model of the design of circuits.

This chapter presents the structure and properties of commonly used algebraic objects, which have many different applications in a variety of areas such as counting techniques, coding theory, information theory, engineering, and circuit design.

## 7.1  Basic Concepts

We start with the basic definitions of operations and their properties.

**DEFINITION 7.1.1**  An ***n*-ary operation** on a set $S$ is a function $* : S \times S \times \cdots \times S \to S$, where the domain is the product of $n$ factors; A **binary operation** on a set $S$ is a function $* : S \times S \to S$; A **monadic operation** (or **unary operation**) on a set $S$ is a function $* : S \to S$.

**DEFINITION 7.1.2**  An **algebraic structure** $(S, *_1, *_2, \cdots, *_k)$ consists of a nonempty set $S$ (the **domain**) and one or more $n$-ary operations $*_i$ defined on $S$.

# 7 Algebra Structures

**EXAMPLE 7.1.1**  $(\mathbf{Z}^+, +)$ is an algebraic structure where $\mathbf{Z}^+$ is the set of all positive integers as defined in Section 3.1, and " $+$ " is the normal addition. $(\mathbf{Z}^+, +, \times)$ is also an algebraic structure where " $\times$ " is the normal multiplication.

Binary operations are the most used operations in algebra. Among binary operations we are particularly interested in those that have special properties. If a binary operation " $*$ " satisfies

$$a*(b*c) = (a*b)*c \quad \text{for all } a,b,c \in S$$

then it is **associative**. If

$$a*b = b*a \quad \text{for all } a,b \in S$$

then " $*$ " is **commutative**. If there is an element $e \in S$ such that

$$e*a = a*e = a \quad \text{for all } a \in S$$

then $e$ is called an **identity** for $S$. If for an element $a \in S$, there is an element $a' \in S$ such that

$$a'*a = a*a' = e$$

then $a'$ is an **inverse** of $a$.

If $a*a = a$ for $a \in S$ then $a$ is an **idempotent element**. If every element is an idempotent element, then " $*$ " is **idempotent**.

**EXAMPLE 7.1.2**  In the algebraic structure $(\mathbf{Z}^+, +)$ the operation " $+$ " is associative and commutative. But there is no identity. In the algebraic structure $(\mathbf{Z}^+, +, \times)$ both operations " $+$ " and " $\times$ " are associative and commutative. For the operation " $\times$ ", the identity is 1. For any $a \in \mathbf{Z}^+$, if $a \neq 1$ then $a$ has no inverse. The inverse of 1 is 1 itself. 1 is the only idempotent element.

A binary operation on a finite set can be represented by a table, as shown in the following example.

**EXAMPLE 7.1.3**  Let $U = \{1,2\}$ and $R = P(U)$ is the power set of $U$ defined in Section 3.1. Define the operations " $+$ " and " $\cdot$ " on the elements of $R$ by

$$A + B = A \triangle B = \{x \mid x \in A \text{ or } x \in B, \text{but not both}\}$$

$$A \cdot B = A \cap B$$

It is easy to see that both operations " $+$ " and " $\cdot$ " are associative. We form two tables for these operations.

**Table 7.1.1  The table for the operation " + " in EXAMPLE 7.1.3**

| +(Δ) | ∅   | {1} | {2} | U   |
|------|-----|-----|-----|-----|
| ∅    | ∅   | {1} | {2} | U   |
| {1}  | {1} | ∅   | U   | {2} |
| {2}  | {2} | U   | ∅   | {1} |
| U    | U   | {2} | {1} | ∅   |

**Table 7.1.2  The table for the operation " · " in EXAMPLE 7.1.3**

| ·(∩) | ∅ | {1} | {2} | U   |
|------|---|-----|-----|-----|
| ∅    | ∅ | ∅   | ∅   | ∅   |
| {1}  | ∅ | {1} | ∅   | {1} |
| {2}  | ∅ | ∅   | {2} | {2} |
| U    | ∅ | {1} | {2} | U   |

Since both tables are symmetric about the diagonal from the upper left to the lower right, both operations described by the tables are commutative. In addition, Table 7.1.1 shows that $\emptyset$ is the additive identity. For any $x \in R$, the additive inverse of $x$ is $x$ itself. Table 7.1.2 shows that $U$ is the multiplicative identity, which is the only element of $P(U)$ having multiplicative inverse.

Using binary operations, we can define some basic algebraic structures whose operations have certain properties.

**DEFINITION 7.1.3**  A **semigroup** $(S, *)$ consists of a nonempty set $S$ and an associative binary operation " $*$ " on $S$. A nonempty subset $T$ of a semigroup $(S, *)$ is a **subsemigroup** of $S$ if $T$ is closed under " $*$ ".

**DEFINITION 7.1.4**  A **monoid** $(S, *)$ consists of a nonempty set $S$ and an associative binary operation " $*$ " on $S$ such that $S$ has an identity. A subset $T$ of a monoid $(S, *)$ with identity $e$ is a submonoid of $S$ if $T$ is closed under " $*$ " and $e \in T$.

**DEFINITION 7.1.5**  Two semigroups [monoids] $(S_1, *_1)$ and $(S_2, *_2)$ are isomorphic if there is a function $\phi: S_1 \to S_2$ that is one-to-one, onto $S_2$, and such that $\phi(a *_1 b) = \phi(a) *_2 \phi(b)$ for all $a, b \in S_1$.

In fact, if two algebraic structures are isomorphic, they are identical from the point of view of algebra.

From DEFINITION 7.1.3 and DEFINITION 7.1.4, it is obvious that every monoid is a semigroup.

**EXAMPLE 7.1.4** $(\mathbf{Z}^+, +)$ is a semigroup but not a monoid. $(\mathbf{Z}^+, \times)$ is a monoid.

We end this section with a few more examples.

**EXAMPLE 7.1.5** The set of closed walks based at a fixed vertex $v$ in a graph forms a monoid under the operation of concatenation. The null walk is the identity.

**EXAMPLE 7.1.6** Semigroup and monoid of transformations on a set $S$: Let $S$ be a nonempty the set and let $F$ be the set of all functions $f: S \to S$. With the operation "$*$" defined by the composition $(f*g)(x) = f(g(x))$, $(F, *)$ is the semigroup of transformations on $S$. The identity of $F$ is the identity transformation $e: S \to S$ where $e(x) = x$ for all $x \in S$. Therefore $(F, *)$ is also the monoid of transformations on $S$.

**EXAMPLE 7.1.7** Let $Z_n = \{0, 1, 2, \cdots, n-1\}$, $n \in \mathbf{Z}^+$. Suppose "$*$" is the binary operation on $\mathbf{Z}_n$ such that $a * b = c$ where $c = (a+b) \mod n$. Then $(\mathbf{Z}_n, *)$ is a monoid.

## WORDS AND EXPRESSIONS

| | |
|---|---|
| Algebraic structures | 代数结构 |
| Monadic operation, or Unary operation | 一元运算 |
| Binary operation | 二元运算 |
| $n$-ary operation | $n$ 元运算 |
| Semigroup | 半群 |
| Monoid | 幺半群,独异点 |
| Idempotent element | 幂等元 |

## EXERCISES 7.1

1. Prove that the following sets are all semigroups and monoids.

   $\mathbf{N} = \{0, 1, 2, 3, \cdots\}$　　　　　　　　　(non-negative integers),
   $\mathbf{Z} = \{\cdots, -2, -1, 0, 1, 2, \cdots\}$　　　　　(integers),
   $\mathbf{Q}$　　　　　　　　　　　　　　　　　(the set of rational numbers),

**R**   (the set of real numbers),
**C**   (the set of complex numbers),

where the operation is either addition or multiplication.

2. Using either addition or multiplication, prove that each monoid in Exercise 1 is a submonoid of each of those following it in the list. For example, ($\mathbf{Q}$, +) is a submonoid of ($\mathbf{R}$, +) and ($\mathbf{C}$, +). Under addition, $e = 0$; under multiplication, $e = 1$.

3. Suppose $(S, *)$ is a monoid with finite elements. Prove that in the operation table of "$*$", any two rows are different.

## 7.2  Groups

Group is the most basic concept of modern algebra. It has wide application in various fields. Although, as a monoid, a group also consists of a set and a binary operation, its binary operation has a stronger nature.

**DEFINITION 7.2.1**   A **group** $(G, *)$ consists of a set $G$ with a binary operation "$*$" defined on $G$ such that

1. "$*$" is associative;
2. $G$ has an identity;
3. for each element $a \in G$ there is an element $a^{-1} \in G$ (inverse of $a$).

If "$*$" is commutative, the group $G$ is **commutative** or **abelian**, named to honor the Norwegian mathematician Niels Henrik Abel.

**EXAMPLE 7.2.1**   $(\mathbf{Z}_n, *)$ defined in EXAMPLE 7.1.7 is an abelian group. This group is normally written as $(\mathbf{Z}_n, +)$. The identity is 0. For any $a \in \mathbf{Z}_n$, the inverse of $a$ is $n - a$ if $a \neq 0$. The inverse of 0 is 0.

**DEFINITION 7.2.2**   The **order** of a **finite group** $G$, denoted as $|G|$, is the number of elements in the group. The group $G$ is **finitely generated** if there are $a_1, a_2, \cdots, a_n \in G$ such that every element of $G$ can be written as $a_{k_1}^{\varepsilon_1}, a_{k_2}^{\varepsilon_2}, \cdots, a_{k_j}^{\varepsilon_j}$ for some $j \geq 0$, where $k_i \in \{1, \cdots, n\}$ and $\varepsilon_i \in \{1, -1\}$, $1 \leq i \leq j$; where the empty product is defined to be $e$.

The operation " $*$ " can be named as multiplication or addition. Conventionally the operation is called multiplication, and the group is written as $(G, \cdot)$. If the group $G$ is abelian, then the operation is usually called addition, and the group is written as $(G, +)$. The following table summarizes the notation conventions for additive groups and multiplicative groups.

**Table 7.2.1 The notation conventions for additive groups and multiplicative groups**

|  | Operation $*$ | Identity $e$ | Inverse $a^{-1}$ |
| --- | --- | --- | --- |
| Additive groups | $a + b$ | 0 | $-a$ |
| Multiplicative groups | $a \cdot b$ or $ab$ | 1 or $e$ | $a^{-1}$ |

From DEFINITION 7.2.1, it is easy to derive the following facts.

**THEOREM 7.2.1**  Every group has exactly one identity.

**THEOREM 7.2.2**  Suppose $G$ is a group, $a \in G$. Then $a$ has exactly one inverse $a^{-1} \in G$. Also we have $(a^{-1})^{-1} = a$. Furthermore, if $a, b \in G$, then $(ab)^{-1} = b^{-1}a^{-1}$. More generally, $(a_1 a_2 \cdots a_k)^{-1} = a_k^{-1} a_{k-1}^{-1} \cdots a_1^{-1}$.

**THEOREM 7.2.3**  If $a$ and $b$ are elements of a group $G$, the equations $ax = b$ and $xa = b$ have unique solutions in $G$, which are $x = a^{-1}b$ and $x = ba^{-1}$, respectively.

**THEOREM 7.2.4**  (Cancellation laws) If $G$ is a group and $a, b, c \in G$ then we have the following:
   **left cancellation law**: if $ab = ac$ then $b = c$;
   **right cancellation law**: if $ba = ca$ then $b = c$.

The following are examples of groups.

**EXAMPLE 7.2.2**  $\mathbf{Z, Q, R, C}$ are groups with the operation of normal addition. The identity is 0. For any element $a$, the inverse of $a$ is $-a$.

**EXAMPLE 7.2.3**  $\mathbf{Q} - \{0\}, \mathbf{R} - \{0\}, \mathbf{C} - \{0\}$ are groups with the operation of normal multiplication. The identity is 1. For any element $a$, the inverse of $a$ is $1/a$.

**EXAMPLE 7.2.4**  Define $M_{m \times n} = \{$all $m \times n$ matrices with entries in $\mathbf{R}\}$ where $m, n \in \mathbf{Z}^+$. Then $M_{m \times n}$ is a group with the operation of normal matrix addition. The identity is the zero matrix $0_{m \times n}$. For any $m \times n$ matrix $A$, the inverse of $A$ is $-A$.

**EXAMPLE 7.2.5** Define $GL(n,\mathbf{R}) = \{$all $n \times n$ nonsingular matrices with entries in $\mathbf{R}\}$ where $n \in \mathbf{Z}^+$. Then $GL(n,\mathbf{R})$ is a group with the operation of normal matrix multiplication. The identity is the identity matrix $I_n$. For any $n \times n$ nonsingular matrix $A$, the inverse of $A$ is $A^{-1}$.

Let us now discuss the homomorphism and isomorphism of groups.

**DEFINITION 7.2.3** For groups $G$ and $H$, a function $\phi: G \to H$ such that $\phi(ab) = \phi(a)\phi(b)$ for all $a, b \in G$ is a **homomorphism**. The notation $a\phi$ is sometimes used instead of $\phi(a)$. The **kernel** of $\phi$ is the set $\{g \in G | \phi(g) = e\}$, where $e$ is the identity of group $H$. A function $\psi: G \to H$ is an **isomorphism** from $G$ to $H$ if $\psi$ is a homomorphism that is one to one and onto $H$. In this case $G$ is **isomorphic** to $H$, written as $G \simeq H$. An isomorphism $\psi: G \to G$ is an **automorphism**.

**THEOREM 7.2.5** Isomorphism is an equivalence relation.
*Proof*: If $G$ is a group then Obviously $G \simeq G$. Also if $G \simeq H$, then $H \simeq G$. Finially if $G \simeq H$ and $H \simeq K$, then $G \simeq K$. Hence isomorphism is reflexive, symmetric and transitive.

**DEFINITION 7.2.4** A **subgroup** of a group $(G, *)$ is a subset $H \subseteq G$ such that $(H, *)$ is a group (with the same group operation as in $G$). Write $H \leqslant G$ if $H$ is a subgroup of $G$.

**THEOREM 7.2.6** A subset $H$ of a group $G$ is a subgroup of $G$ if and only if the following are all true:
  1. $H \neq \emptyset$;
  2. $a, b \in H$ implies $ab \in H$;
  3. $a \in H$ implies $a^{-1} \in H$.

*Proof*: We only prove the "if" part. From DEFINITION 7.2.1, we only need to prove that the identity $e$ of $G$ is in $H$. Since $H \neq \emptyset$, there is an element $a \in H$. This implies $a^{-1} \in H$. Consequently $e = aa^{-1} \in H$.

**EXAMPLE 7.2.6** According to DEFINITION 7.2.4, if $G$ is a group, then $G$ and $\{e\}$ are subgroups of $G$. They are **improper subgroups** or **trivial subgroups** of $G$. All other subgroups of $G$ are **proper subgroups** or **nontrivial subgroups** of $G$.

**EXAMPLE 7.2.7** If $G$ is a group and $a \in G$, the set $(a) = \{\cdots, a^{-2} = (a^{-1})^2, a^{-1}, a^0 = e, a, a^2, \cdots\} = \{a^n | n \in \mathbf{Z}\}$ is a subgroup of $G$. It is called the **cyclic group**

generated by $a$. The element $a$ is a **generator** of $G$.

**THEOREM 7.2.7**  If $p$ is a prime then there is only one group of order $p$ (up to isomorphism), namely the group $(\mathbf{Z}_p, +)$.

*Proof*: Since $p$ is prime, it can be proved that a group $G$ of order $p$ is a cyclic group (COROLLARY 7.2.1). Suppose $G = (a) = \{a, a^2, \cdots, a^{p-1}, a^p = e\}$, then the function $f: G \to \mathbf{Z}_p$ defined by
$$f(a^k) = k \text{ if } k < p,$$
$$f(e) = 0$$
is an isomorphism.

**THEOREM 7.2.8**  If $\phi: G \to H$ is a homomorphism, then $\phi(G)$ is a subgroup of $H$. The kernel of $\phi$ is a subgroup of $G$.

The proof of THEOREM 7.2.8 is left to the readers.

**THEOREM 7.2.9**  (**Lagrange**) Let $G$ be a finite group. If $H$ is a subgroup of $G$, then the order of $H$ is a divisor of the order of $G$.

The proof of Lagrange's theorem is out of the range of this book hence omitted.

**COROLLARY 7.2.1**  Any group of prime order is cyclic.

Permutation groups are a kind of groups having fundamental importance in matrix theory.

**DEFINITION 7.2.5**  A **permutation** is a one-to-one and onto function $\sigma: S \to S$, where $S$ is a nonempty set. If $S = \{a_1, a_2, \cdots, a_n\}$, a permutation $\sigma$ is sometimes written as the $2 \times n$ matrix
$$\sigma = \begin{pmatrix} a_1 & a_2 & \cdots & a_n \\ a_1\sigma & a_2\sigma & \cdots & a_n\sigma \end{pmatrix}$$
where $a_i\sigma$ means $\sigma(a_i)$.

**DEFINITION 7.2.6**  A permutation $\sigma: S \to S$ is a **cycle** of length $n$ if there is a subset of $S$ of size $n$, $\{a_1, a_2, \cdots, a_n\}$, such that $a_1\sigma = a_2, a_2\sigma = a_3, \cdots, a_n\sigma = a_1$, and $a\sigma = a$ for all other elements of $S$. Write $\sigma = (a_1 a_2 \cdots a_n)$. A **transposition** is a cycle of length 2.

**EXAMPLE 7.2.8**  A **permutation group** $(G, X)$ is a collection $G$ of permutations on a nonempty set $X$ (whose elements are called objects) such that these permutations form a group under composition. That is, if $\sigma$ and $\tau$ are permutations in $G$, $\sigma\tau$ is the permutation in $G$ defined by the rule $a(\sigma\tau) = (a\sigma)\tau$. The **order** of the permutation group is $|G|$. The **degree** of the permutation group is $|X|$. If $|X| = n$, the group of permutations is written as $S_n$, called the **symmetric group** of degree $n$.

**DEFINITION 7.2.7** An **isomorphism** from a permutation group $(G, X)$ to a permutation group $(H, Y)$ is a pair of functions $(\alpha: G \to H, f: X \to Y)$ such that $\alpha$ is a group isomorphism and $f$ is one-to-one and onto $Y$.

**DEFINITION 7.2.8** An **even permutation** [**odd permutation**] is a permutation that can be written as a product of an even [odd] number of transpositions. The **sign** of a permutation (where the permutation is written as a product of transpositions) is $+1$ if it has an even number of transpositions and $-1$ if it has an odd number of transpositions. The **identity** permutation on $S$ is the permutation $\iota: S \to S$ such that $x\iota = x$ for all $x \in S$.

The following Cayley's theorem plays an important part in the group theory, which characterizes the structures of finite groups. The proof of Cayley's theorem is out of the range of this book and so omitted.

**THEOREM 7.2.10** (**Cayley**) If $G$ is a finite group of order $n$, then $G$ is isomorphic to a subgroup of $S_n$.

The groups of rigid motions are also a very important kind of groups. We list two typical examples.

**EXAMPLE 7.2.9** The **dihedral group** (**octic group**) $D_n$ is the group of rigid motions (rotations and reflections) of a regular polygon with $n$ sides under composition.

**EXAMPLE 7.2.10** The **Klein four-group** (or **Viergruppe** or the **group of the rectangle**) is the group under composition of the four rigid motions of a rectangle that leave the rectangle in its original location. The adjective Klein honors the mathematician Felix Klein. (The Klein four-group consists of the following four rigid motions of a rectangle: the rotations about the center through $0°$ or $180°$, and reflections through the horizontal or vertical lines through its center.)

## WORDS AND EXPRESSIONS

| | |
|---|---|
| Group | 群 |
| Finite group | 有限群 |

| Abelian group | 阿贝尔群,交换群 |
| Subgroup | 子群 |
| Trivial subgroup | 平凡子群 |
| Proper subgroup | 真子群 |
| Cyclic group | 循环群 |
| Permutation group | 置换群 |
| Klein four-group | 克莱因四元群 |
| Automorphism | 自同构 |

## EXERCISES 7.2

1. Prove that for any group $G$, the identity of $G$ is unique. The inverse of each element of $G$ is unique.

2. Prove that the monoids in Exercise 1 of Section 7.1 are groups.

3. Prove that if $H$ is a subgroup of a group $G$, then the identity element of $H$ is the identity element of $G$; the inverse (in the subgroup $H$) of an element $a$ in $H$ is the inverse (in the group $G$) of $a$.

4. A subset $H$ of a group $G$ is a subgroup of $G$ if and only if $H \neq \emptyset$ and $a, b \in H$ implies that $ab^{-1} \in H$.

5. Prove the "only if" part of THEOREM 7.2.6.

6. Find examples to show that the multiplication operation of permutations is not commutative.

7. Suppose $K = \{e, a, b, c\}$, and define an operation $*$ on $K$ by the following table.

| $*$ | $e$ | $a$ | $b$ | $c$ |
|---|---|---|---|---|
| $e$ | $e$ | $a$ | $b$ | $c$ |
| $a$ | $a$ | $e$ | $c$ | $b$ |
| $b$ | $b$ | $c$ | $e$ | $a$ |
| $c$ | $c$ | $b$ | $a$ | $e$ |

Prove that $(K, *)$ is a group, and $(K, *)$ is isomorphic to Klein four-group.

## 7.3 Rings and Fields

In last section we studied the algebraic structures with one binary operation. Now we turn to the algebraic structures with two binary operations.

**DEFINITION 7.3.1** A **ring** $(R, +, \cdot)$ consists of a set $R$ closed under binary operations "$+$" (**addition**) and "$\cdot$" (**multiplication**) such that:
1. $(R, +)$ is an abelian group; i.e., $(R, +)$ satisfies:
   a. associative property: $a + (b + c) = (a + b) + c$ for all $a, b, c \in R$;
   b. identity property: $R$ has an identity element, $0$, that satisfies $0 + a = a + 0 = a$ for all $a \in R$;
   c. inverse property: for each $a \in R$ there is an additive inverse element $-a \in R$ (the negative of $a$) such that $-a + a = a + (-a) = 0$;
   d. commutative law: $a + b = b + a$ for all $a, b \in R$;
2. $(R, \cdot)$ is a semigroup; i.e., the operation $\cdot$ is associative: $a \cdot (b \cdot c) = (a \cdot b) \cdot c$ for all $a, b, c \in R$;
3. the **distributive properties** for multiplication over addition hold for all $a, b, c \in R$:
   a. left distributive property: $a \cdot (b + c) = a \cdot b + a \cdot c$;
   b. right distributive property: $(a + b) \cdot c = a \cdot c + b \cdot c$.

A ring $R$ is **commutative** if the multiplication operation is commutative: $a \cdot b = b \cdot a$ for all $a, b \in R$.

A ring $R$ is a **ring with unity** if there is an identity, $1 (\neq 0)$, for multiplication; i.e., $1 \cdot a = a \cdot 1 = a$ for all $a \in R$. The multiplicative identity is the **unity** of $R$.

An element $x$ in a ring $R$ with unity is a **unit** if $x$ has a multiplicative inverse; i.e., there is $x^{-1} \in R$ such that $x \cdot x^{-1} = x^{-1} \cdot x = 1$.

**Subtraction** in a ring is defined by the rule $a - b = a + (-b)$.

Multiplication, $a \cdot b$, is often written as $ab$ or $a \times b$. The order of precedence of operations in a ring follows that for real numbers: multiplication is to be done before addition. That is, $a + bc$ means $a + (bc)$ rather than $(a + b)c$.

**EXAMPLE 7.3.1** $(\mathbf{Z}, +, \cdot)$, $(\mathbf{Q}, +, \cdot)$, $(\mathbf{R}, +, \cdot)$ and $(\mathbf{C}, +, \cdot)$ are all rings where "$+$" and "$\cdot$" are usual addition and multiplication of numbers, respectively.

**EXAMPLE 7.3.2** (**Polynomial rings**) For a ring $R$, the set
$$R[x] = \{a_n x^n + \cdots + a_1 x + a_0 \mid a_0, a_1, \cdots, a_n \in R\}$$

forms a ring, where the elements are added and multiplied using the usual rules for addition and multiplication of polynomials. The additive identity, $0$, is the constant polynomial $p(x)=0$; the unity is the constant polynomial $p(x)=1$ if $R$ has a unity $1$.

**EXAMPLE 7.3.3** The set of all units of a ring is a group under the multiplication defined on the ring.

**EXAMPLE 7.3.4** Suppose $S$ is a set. In Chapter 3 we introduced $P(S)$ as the power set of $S$. Define the operations " $+$ " and " $\cdot$ " on the elements of $P(S)$ as following: for any $A, B \in P(S)$

$$A + B = A \triangle B = \{x \mid x \in A \text{ or } x \in B, \text{but not both}\}$$
$$A \cdot B = A \cap B$$

Then $(P(S), +, \cdot)$ is a ring, called the **subset ring** of $S$. Since $\cap$ is commutative, and $(P(S), \cdot)$ has the identity $S$, $(P(S), +, \cdot)$ is a commutative ring with unity $S$.

**THEOREM 7.3.1** Suppose $(R, +, \cdot)$ is a ring. For any $a, b, c \in R$

1. $a0 = 0a = 0$
2. $a(-b) = (-a)b = -(ab)$
3. $(-a)(-b) = ab$
4. $a(b-c) = ab - ac$
5. $(a-b)c = ac - bc$
6. if the ring has unity, then $(-1)a = -a$

*Proof*: We only prove the first one, and leave the rest of the proof to the readers. Since

$$0a = (0+0)a = 0a + 0a,$$

according to the cancellation laws (THEOREM 7.2.4) we have

$$0a = 0.$$

$a0 = 0$ can be proved in the same way.

**DEFINITION 7.3.2** A subset $S$ of a ring $(R, +, \cdot)$ is a **subring** of $R$ if $(S, +, \cdot)$ is a ring using the same operations $+$ and $\cdot$ that are used in $R$. A subset $I$ of a ring $(R, +, \cdot)$ is an **ideal** of $R$ if:

1. $(I, +, \cdot)$ is a subring of $(R, +, \cdot)$;
2. $I$ is closed under left and right multiplication by elements of $R$: if $x \in I$ and $r \in R$, then $rx \in I$ and $xr \in I$.

In a commutative ring $R$, an ideal $I$ is **principal** if there is $r \in R$ such that $I = Rr = \{xr \mid x \in R\}$. $I$ is the **principal ideal generated by** $r$, written as $I = (r)$.

In a commutative ring $R$, an ideal $I \neq R$ is **maximal** if the only ideal properly containing $I$ is $R$.

In a commutative ring $R$, an ideal $I \neq R$ is **prime** if $ab \in I$ implies that $a \in I$ or $b \in I$.

**THEOREM 7.3.2** Suppose $R$ is a ring, $S$ is a nonempty set of $R$. If
1. $a - b \in S$ for any $a, b \in S$;
2. $ab \in S$ for any $a, b \in S$

then $S$ is a subring of $R$.

*Proof*: From condition 1, $(S, +)$ is a commutative subgroup of $(R, +)$. From condition 2, $(S, \cdot)$ is a subsemigroup of $(R, \cdot)$. Obviously the distributive properties for multiplication over addition hold for all $a, b, c \in S$. According to DEFINITION 7.3.1 and DEFINITION 7.3.2, $(S, +, \cdot)$ is a subring of $(R, +, \cdot)$.

**EXAMPLE 7.3.5** Let $n$ be a positive integer. Define
$$n\mathbf{Z} = \{nz \mid z \in \mathbf{Z}\}$$
Then $n\mathbf{Z}$ is a nonempty subset of $\mathbf{Z}$. Using THEOREM 7.3.2, it is easy to see that $(n\mathbf{Z}, +, \cdot)$ is a subring of $(\mathbf{Z}, +, \cdot)$. It can be proved that $n\mathbf{Z}$ is a principal ideal of $\mathbf{Z}$ generated by $n$.

**DEFINITION 7.3.3** If $R$ and $S$ are rings, a function $\phi: R \to S$ is a **ring homomorphism** if for all $a, b \in R$:

$\phi(a + b) = \phi(a) + \phi(b)$      ($\phi$ preserves addition)
$\phi(ab) = \phi(a)\phi(b)$.      ($\phi$ preserves multiplication)

If a ring homomorphism $\phi$ is also one-to-one and onto $S$, then $\phi$ is a **ring isomorphism** and $R$ and $S$ are **isomorphic**, written as $R \cong S$.

A ring **endomorphism** is a ring homomorphism $\phi: R \to R$.

A ring **automorphism** is a ring isomorphism $\phi: R \to R$.

The **kernel** of a ring homomorphism $\phi: R \to S$ is $\phi^{-1}(0) = \{x \in R \mid \phi(x) = 0\}$.

**EXAMPLE 7.3.6** The function $\phi: Z \to Z_n$ defined by the rule $\phi(a) = a \bmod n$ is a ring homomorphism.

**THEOREM 7.3.3** If $\phi: R \to S$ is a ring homomorphism, $\phi(R)$ is a subring of $S$. If $\phi: R \to S$ is a ring homomorphism and $R$ has unity, either $\phi(1) = 0$ or $\phi(1)$ is the unity

for $\phi(R)$. The kernel of a ring homomorphism from $R$ to $S$ is an ideal of the ring $R$.
*Proof*: it is a corollary of Exercise 1 and Exercise 2 in EXERCISES 7.3.

Beginning with rings, as additional requirements are added, a number of algebraic structures can be obtained.

**DEFINITION 7.3.4**  The **cancellation properties** in a ring $R$ state that for all $a, b, c \in R$:

   if $ab = ac$ and $a \neq 0$, then $b = c$     (left cancellation property)

   if $ba = ca$ and $a \neq 0$, then $b = c$     (right cancellation property).

Let $R$ be a ring and $a, b \in R$ where $a \neq 0$, $b \neq 0$. If $ab = 0$, then $a$ is a **left divisor** of zero and $b$ is a **right divisor** of zero.

**DEFINITION 7.3.5**  An **integral domain** is a commutative ring with unity that has no zero divisors.

**EXAMPLE 7.3.7**  $(\mathbf{Z}, +, \cdot)$, $(\mathbf{Q}, +, \cdot)$, $(\mathbf{R}, +, \cdot)$ and $(\mathbf{C}, +, \cdot)$ are all integral domains.

**EXAMPLE 7.3.8**  The rings $n\mathbf{Z}$ defined in EXAMPLE 7.3.5 are not integral domains when $n > 1$ since they contains no unity.

**THEOREM 7.3.4**  Let $(R, +, \cdot)$ be a ring. $R$ has no zero divisors if and only the operation "$\cdot$" has the cancellation properties, i.e., for any $a, b, c \in R, a \neq 0$,

   if $ab = ac$, then $b = c$;

   if $ba = ca$, then $b = c$.

*Proof*: Sufficiency: For any $a, b \in R$, if $ab = 0$ and $a \neq 0$, then $ab = 0 = a0$. From cancellation properties we have $b = 0$, which means that $R$ has no zero divisors.

Necessity: For any $a, b, c \in R, a \neq 0$, from $ab = ac$ we have $a(b - c) = 0$. Since $R$ has no zero divisors and $a \neq 0$ we have $b - c = 0$, i.e., $b = c$. This proves that "$\cdot$" has the left cancellation property. Similarly we can prove the right cancellation property of "$\cdot$".

**DEFINITION 7.3.6**  A field $(F, +, \cdot)$ consists of a set $F$ together with two binary operations, "$+$" and "$\cdot$", such that:

   1. $(F, +, \cdot)$ is a ring;

   2. $(F - \{0\}, \cdot)$ is a commutative group.

Obviously every field is a commutative ring with unity since a field satisfies all properties of a commutative ring with unity, and has the additional property that every nonzero element has a multiplicative inverse.

**EXAMPLE 7.3.9** $(\mathbf{Q}, +, \cdot)$, $(\mathbf{R}, +, \cdot)$ and $(\mathbf{C}, +, \cdot)$ are all fields. $(\mathbf{Z}, +, \cdot)$ is an integral domain but is not a field, since for any $z \in \mathbf{Z}$, if $z > 1$ then $z$ has no inverse in $\mathbf{Z}$.

**THEOREM 7.3.5** Every finite integral domain is a field.

*Proof*: Suppose $(A, +, \cdot)$ is a finite integral domain. For any $a, b, c \in A$ and $c \neq 0$, if $a \neq b$ then $ac \neq bc$. Since $A$ is closed under the operation " $\cdot$ ", we have $A \cdot c = A$. Therefore, for the unity 1 of $A$, there is $d \in A$ such that $d \cdot c = 1$, which means that $d$ is the inverse of $c$. Consequently $A$ is a field.

**DEFINITION 7.3.7** A subfield $F$ of field $(K, +, \cdot)$ is a subset of $K$ that is a field using the same operations as those in $K$.

**EXAMPLE 7.3.10** The intersection of all subfields of a field $F$ is a field, called the **prime field** of $F$.

**DEFINITION 7.3.8** A field isomorphism is a function $\phi: F_1 \to F_2$, where $F_1$ and $F_2$ are fields, such that $\phi$ is one-to-one, onto $F_2$, and satisfies the following for all $a, b \in F_1$:

$$\phi(a+b) = \phi(a) + \phi(b);$$
$$\phi(ab) = \phi(a)\phi(b).$$

A field automorphism is an isomorphism $\phi: F \to F$, where $F$ is a field. The set of all automorphisms of $F$ is denoted by $\mathrm{Aut}(F)$.

**EXAMPLE 7.3.11** If $F$ is a field, $\mathrm{Aut}(F)$ is a group under composition of functions.

## WORDS AND EXPRESSIONS

| | |
|---|---|
| Ring | 环 |
| Ring with unity | 含幺环 |
| Polynomial rings | 多项式环 |
| Subring | 子环 |
| Ideal | 理想 |
| Ring homomorphism | 环同态 |
| Endomorphism | 自同态 |
| Kernel | 核 |

Integral domain  整环
Field  域

## EXERCISES 7.3

1. Prove that homomorphisms preserve subrings: Let $\phi:R\to S$ be a ring homomorphism. If $A$ is a subring of $R$, then $\phi(A)$ is a subring of $S$. If $B$ is a subring of $S$, then $\phi^{-1}(B)$ is a subring of $R$.
2. Prove that homomorphisms preserve ideals: Let $\phi:R\to S$ be a ring homomorphism. If $A$ is an ideal of $R$, then $\phi(A)$ is an ideal of $S$. If $B$ is an ideal of $S$, then $\phi^{-1}(B)$ is an ideal of $R$.
3. Prove that $\mathbf{Z}_n$ is an integral domain if and only if $n$ is a prime.
4. Prove that if $F$ is a field and $a,b \in F$ where $a \neq 0$, then $ax+b=0$ has a unique solution in $F$.

## 7.4  Boolean Algebras

In preceding sections we studied a number of algebraic structures with one or two binary operations. In this section we will learn a new kind of algebraic structures with two binary operations and one unary operation—Boolean algebra.

Boolean algebra was named after the English Mathematician George Boole who, in 1854, published the book "An investigation into laws of thought", which set up the fundamental logical methodology. Boole showed that logical propositions and their connectives can be expressed in the language of set theory. For this reason Boolean algebra is sometimes referred to as the algebra of logic.

By taking variables to represent values of "on" and "off", Boolean algebra is used to design and analyze digital switching circuitry, which is the base of the world of electronics producing products such as computers, mobile phones, CD players, and so on. In this section we briefly introduce the basic idea of Boolean algebras, and show some examples of its application in circuit technology.

**DEFINITION 7.4.1**  A Boolean algebra $(B, +, \cdot, ^-, 0, 1)$ consists of a set $B$ closed under two binary operations, " + " (**addition**) and " $\cdot$ " (**multiplication**), and one monadic operation, " $^-$ " (**complementation**), and having two distinct elements, 0 and 1, such that the following laws are true for all $a,b,c \in B$:

1. **commutative laws**:
   a. $a + b = b + a$;
   b. $a \cdot b = b \cdot a$.
2. **distributive laws**:
   a. $a \cdot (b + c) = (a \cdot b) + (a \cdot c)$;
   b. $a + (b \cdot c) = (a + b) \cdot (a + c)$.
3. **identity laws**:
   a. $a + 0 = a$;
   b. $a \cdot 1 = a$.
4. **complement laws**:
   a. $a + \overline{a} = 1$;
   b. $a \cdot \overline{a} = 0$.

Sometimes we also call the set $B$ a Boolean algebra if we do not need to talk about the operations. In computer science, the operations defined in DEFINITION 7.4.1 are commonly called **AND**, **OR** and **NOT**. Also, the two values 0 and 1 are referred to as **Boolean values**, and sometimes written as **FALSE** and **TRUE**. The following two examples give a good reason for the naming, which also show that Boolean algebra is a natural extension of propositional logic.

**EXAMPLE 7.4.1** Let $\Gamma$ be the set of all propositions, then $(\Gamma, \vee, \wedge, \neg, F, T)$ is a Boolean algebra, where, " $\vee$ ", " $\wedge$ " and " $\neg$ " are logical operators disjunction (or), conjunction (and) and negation (not) defined in Section 1.1, respectively. $F$ is the proposition that is always false, and $T$ is the proposition that is always true. This conclusion can be easily obtained from the Logical Basic Equivalences stated in Section 1.3.

**EXAMPLE 7.4.2** Given $n$ propositional variables, the set of all wffs in these variables (identified with their truth tables) is a Boolean algebra where

$$A + B = A \vee B, A \cdot B = A \wedge B, \overline{A} = \neg A$$

and 0 is a contradiction (the truth table with only values $F$) and 1 is a tautology (the truth table with only values $T$).

The following example shows a basic model for finite Boolean algebra which we will make use of later. Note that the definitions of " $+$ " and " $\cdot$ " in the example are actually same as Boolean operations " $\vee$ " and " $\wedge$ ".

## 7 Algebra Structures

**EXAMPLE 7.4.3** Denote $Z_2 = \{0,1\}$. $(Z_2, +, \cdot, ^-, 0, 1)$ is a Boolean algebra, where addition, multiplication and complementation are defined by the following tables:

**Table 7.4.1  Operations for $Z_2$**

| + | 0 | 1 |
|---|---|---|
| 0 | 0 | 1 |
| 1 | 1 | 1 |

| $\cdot$ | 0 | 1 |
|---|---|---|
| 0 | 0 | 0 |
| 1 | 0 | 1 |

| $x$ | $\bar{x}$ |
|---|---|
| 0 | 1 |
| 1 | 0 |

**EXAMPLE 7.4.4** Suppose $S$ is a set. Recall that we defined the power set $P(S)$ of $S$ in Chapter 3 as the collection of all subsets of $S$. It is easy to verify that $(P(S), \cup, \cap, ^-, S, \phi)$ is a Boolean algebra.

It is common practice to omit the "$\cdot$" symbol in a Boolean algebra, writing $ab$ instead of $a \cdot b$. By convention, complementation is done first, then multiplication, and finally addition. For example, $a + b\bar{c}$ means $a + (b(\bar{c}))$.

**THEOREM 7.4.1** Suppose that $(B, +, \cdot, ^-, 0, 1)$ is a Boolean algebra, then the following laws are true for all $a, b, c \in B$:

1. **associative laws:** $a + (b + c) = (a + b) + c$, $a(bc) = (ab)c$ ( Hence there is no ambiguity in writing $a + b + c$ and $abc$. )
2. **idempotent laws:** $a + a = a$, $aa = a$
3. **absorption laws:** $a(a + b) = aa + ab = a$
4. **domination (boundedness) laws:** $a + 1 = 1$, $a0 = 0$
5. **double complement (involution) law:** $\bar{\bar{a}} = a$
6. **De Morgan's laws:** $\overline{(a + b)} = \bar{a}\bar{b}$, $\overline{(ab)} = \bar{a} + \bar{b}$
7. **uniqueness of complement:** if $a + b = 1$ and $ab = 0$, then $b = \bar{a}$.

*Proof*: The proof is similar to those for the Logical Basic Equivalences in Section 1.3. It is left for the readers as an exercise.

**DEFINITION 7.4.2** The **dual** of a statement in a Boolean algebra is the statement obtained by interchanging the operations "$+$" and "$\cdot$" and interchanging the elements 0 and 1 in the original statement. An element $a \neq 0$ in a Boolean algebra $B$ is an **atom** if the following holds: for any $x \in B$ if $xa = x$, then either $x = 0$ or $x = a$.

The following result is a very useful tool in Boolean algebra. We state it without proof.

**THEOREM 7.4.2 ( Duality principle )** If a theorem is the consequence of the definition of Boolean algebra, then the dual of the theorem is also a theorem.

**DEFINITION 7.4.3**  Two Boolean algebras $B_1$ and $B_2$ are **isomorphic** ( as Boolean algebras ) if there is a function $\phi: B_1 \to B_2$ that is one-to-one and onto $B_2$ such that for all $a, b \in B1$:
1. $\phi(a+b) = \phi(a) + \phi(b)$;
2. $\phi(ab) = \phi(a)\phi(b)$;
3. $\phi(\overline{a}) = \overline{\phi(a)}$.

Given a Boolean algebra $(B, +, \cdot, ^-, 0, 1)$, dose $B$ have a partial order? The following theorem confirm the answer by defining such an order based on " $+$ ".

**THEOREM 7.4.3**  Given a Boolean algebra $(B, +, \cdot, ^-, 0, 1)$, $B$ is a partially ordered set.

*Proof*: For any $a, b \in B$, define $a \leqslant b$ if and only if $a + b = b$. It is easy to see that this relation on $B$ is reflexive, antisymmetric, and transitive.

**EXAMPLE 7.4.5**  An element $a \neq 0$ in a Boolean algebra is an atom if the following holds: if $x \leqslant a$, then either $x = 0$ or $x = a$.

*Proof*: According to absorption laws, $xa = x$ is equivalent to $x + a = a$. Therefore if for any $x \in B$, $xa = x$ then $x + a = a$. By THEOREM 7.4.3 we have $x \leqslant a$. Consequently either $x = 0$ or $x = a$. From DEFINITION 7.4.2, $a$ is an atom.

We defined **NAND** and **NOR** for propositions in Section 1.5. As a fact that Boolean algebra is a generalization of the algebra of logical propositions, the following definition generalizes these concepts to Boolean algebra.

**DEFINITION 7.4.4**  Let $(B, +, \cdot, ^-, 0, 1)$ be a Boolean algebra, the binary operation **NAND**, **NOR**, and **XOR** on $B$ are defined as follows:
1. NAND, written as $|$, is defined by $a | b = \overline{(ab)}$.
2. NOR, written as $\downarrow$, is defined by $a \downarrow b = \overline{(a+b)}$.
3. XOR, written as $\oplus$, is defined by $a \oplus b = a\overline{b} + \overline{a}b$.

Boolean algebras have simple structures if they are finite, as summarized in the following theorem.

**THEOREM 7.4.4**  Suppose that $B$ is a Boolean algebra with finite elements. Then $B$ is isomorphic to $Z_2^n$ for some positive integer $n$, where $Z_2 = \{0, 1\}$ is defined in

EXAMPLE 7.4.3. Hence every finite Boolean algebra has $2^n$ elements. The atoms of $B$ are the $n$ $n$-tuples of 0s and 1s with a 1 in exactly one position. If $b \in B(b \neq 0)$, there is exactly one set of atoms $a_1, a_2, \cdots, a_k$ such that $b = a_1 + a_2 + \cdots + a_k$.

Conversely, we have the following result.

**THEOREM 7.4.5** If a Boolean algebra $B$ has $n$ atoms, then $B$ has $2^n$ elements.

The proof of THEOREM 7.4.4 and THEOREM 7.4.5 is out of the range of the book hence omitted here.

In Chapter 2 we found that propositional function is a powerful tool for predicate logic. We now generalize the concept to Boolean algebra. Before start, we need a series of concepts in parallel of those in Chapter 2, such as variables, literals, disjunctive normal forms, and conjunctive normal forms. Since the principles are similar, we will only briefly introduce them without detail discussion.

**DEFINITION 7.4.5** A **Boolean variable** is a variable which assumes only Boolean values, i.e., 0 and 1, or FALSE and TRUE. A **literal** is a Boolean variable or the complement of a Boolean variable.

From the definition, a Boolean variable takes values in the set $Z_2$. That is why $Z_2$ is sometimes called **the Boolean domain**. Since $Z_2$ is a Boolean algebra, we can define the addition, multiplication and complements of Boolean variables in the same way as the definition in Table 7.4.1, which is the way we define " $\vee$ ", " $\wedge$ ", and " $\neg$ " in Chapter 1. It is obvious that the difference between the ordinary variables and Boolean variables is that while the ordinary variables assume infinitely values, Boolean variables can only assume two possible values. In the rest of this chapter, all variables referred are Boolean variables if we don't indicate otherwise.

**DEFINITION 7.4.6** Suppose $x_1, \cdots, x_n$ are Boolean variables. A **Boolean expression** in $x_1, \cdots, x_n$ is an expression defined recursively by:
1. 0, 1, and all variables $x_i (i = 1, \cdots, n)$ are Boolean expressions in $x_1, \cdots, x_n$;
2. if $E_1$ and $E_2$ are Boolean expressions in the variables $x_1, \cdots, x_n$, then $E_1 E_2$, $E_1 + E_2$, and $\overline{E_1}$ are Boolean expressions in the variables $x_1, \cdots, x_n$.

**DEFINITION 7.4.7** A **minterm** of Boolean variables $x_1, \cdots, x_n$ is a product of the form $y_1 \cdots y_n$ where for each $i$, $y_i$ is equal to $x_i$ or $\overline{x_i}$. A **maxterm** of $x_1, \cdots, x_n$ is a sum of the form $y_1 + \cdots + y_n$ where for each $i$, $y_i$ is equal to $x_i$ or $\bar{x}_i$.

**DEFINITION 7.4.8** A **Boolean function** of degree $n$ is a function $f: Z_2^n \to Z_2$. A Boolean function is in **disjunctive normal form** (**DNF**) (or **sum-of-products expansion**) if it is written as a sum of distinct minterms in the variables $x_1, \cdots, x_n$. A Boolean function is in **conjunctive normal form** (**CNF**) (or **product-of-sums expansion**) if it is written as a product of distinct maxterms.

**EXAMPLE 7.4.6** Since a Boolean function of degree $n$ can be expressed as a propositional expression in $n$ variables, and two propositional expressions are logically equivalent if and only if they express the same Boolean function, there are $2^{2^n}$ Boolean functions of degree $n$. In the special case of $n = 2$, we have 16 different Boolean functions with two variables, $x$ and $y$, which are given in the following tables.

Table 7.4.2 The 16 Boolean functions of degree 2

| $x$ | $y$ | 1 | $x+y$ | $\overline{x}+y$ | $x+\overline{y}$ | $x\mid y$ | $x$ | $y$ | $x \oplus y$ |
|---|---|---|---|---|---|---|---|---|---|
| 1 | 1 | 1 | 1 | 1 | 1 | 0 | 1 | 1 | 0 |
| 1 | 0 | 1 | 1 | 1 | 0 | 1 | 1 | 0 | 1 |
| 0 | 1 | 1 | 1 | 0 | 1 | 1 | 0 | 1 | 1 |
| 0 | 0 | 1 | 0 | 1 | 1 | 1 | 0 | 0 | 0 |

| $x$ | $y$ | 0 | $x \downarrow y$ | $\overline{x}y$ | $x\overline{y}$ | $xy$ | $\overline{x}$ | $\overline{y}$ | $\overline{x \oplus y}$ |
|---|---|---|---|---|---|---|---|---|---|
| 1 | 1 | 0 | 0 | 0 | 0 | 1 | 0 | 0 | 1 |
| 1 | 0 | 0 | 0 | 0 | 1 | 0 | 0 | 1 | 0 |
| 0 | 1 | 0 | 0 | 1 | 0 | 0 | 1 | 0 | 0 |
| 0 | 0 | 0 | 1 | 0 | 0 | 0 | 1 | 1 | 1 |

**THEOREM 7.4.6** Every Boolean function can be written as a Boolean expression.

**THEOREM 7.4.7** Every Boolean function (not identically 0) can be written in disjunctive normal form. Every Boolean function (not identically 1) can be written in conjunctive normal form.

## 7  Algebra Structures

*Proof*: Suppose that $f$ is a Boolean function other than an identical 0, and $g$ is a Boolean function other than an identical 1.

We can use either of the following 2 ways to write $f$ in disjunctive normal form:
1. Rewrite the expression for $f$ so that no parentheses remain. For each term that does not have a literal for a variable $x_i$, multiply that term by $x_i + \overline{x_i}$. Multiply out so that no parentheses remain. Use the idempotent law to remove any duplicate terms or duplicate factors.
2. Make a table of values for $f$. For each row where $f$ has the value 1, form a minterm that yields 1 in only that row. Form the sum of these minterms.

We can use any of the following 3 ways to write $g$ in conjunctive normal form:
1. Write the negation of $g$ in disjunctive normal form. Then use De Morgan's laws to take the negation of this expression.
2. Make a table of values for $g$. For each row where $g$ has the value 0, form a minterm that yields 1 in only that row. Form the sum of these minterms. Use De Morgan's laws to take the complement of this sum.
3. Make a table of values for $g$. For each row where $g$ has the value 0, form a maxterm that yields 0 in only that row. Form the product of these maxterms.

The concept of functionally complete introduced in Section 1.5 can also be applied to Boolean algebra as shown below.

**DEFINITION 7.4.9**  A set of operators in a Boolean algebra is **functionally complete** if every Boolean function can be written using only these operators.

**EXAMPLE 7.4.7**  The following sets of operators are functionally complete sets, which can be proved in a similar way as in Section 1.5.
1. $\{+, \cdot, ^-\}$.
2. $\{+, ^-\}$.
3. $\{\cdot, ^-\}$.
4. $\{|\}$.
5. $\{\downarrow\}$.

Also we know that the set $\{+, \cdot\}$ is not complete.

Boolean algebra can be used to model circuitry, with 0s and 1s as inputs and outputs. The elements of these circuits are **gates**. The term "gate" is used to describe the members of a set of basic electronic components which are physical realization of the Boolean operations. A group of gates, when combined with each other properly, are able to perform complex logical and arithmetic operations, and so implement Boolean expressions. Electronic circuits which combine digital signals according to Boolean algebra are referred to as logic gates, because they control the flow of information.

There are three standard gates, OR, AND and NOT implementing, " + ", " · " and " $-$ ", respectively, which are depicted in Figure 7.4.1

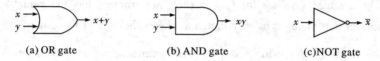

(a) OR gate        (b) AND gate        (c)NOT gate

Figure 7.4.1    Standard OR, AND, and NOT gates.

OR and AND gates can take more than two inputs. Figure 7.4.2 shows an OR gate and an AND gate with multiple inputs. They correspond to $x_1 \cdots x_n$ and $x_1 + \cdots + x_n$.

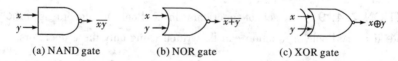

Figure 7.4.2    OR and AND gates taking multiple inputs

The three operations defined in DEFINITION 7.4.4 are also elementary. Their corresponding gates are given in the following figure.

(a) NAND gate        (b) NOR gate        (c) XOR gate

Figure 7.4.3    NAND, NOR, and XOR gates

Circuits are electronic devices which take Boolean values as input, and output Boolean values as results. A Boolean function can be transformed from an algebraic expression into a circuit diagram composed of logic gates. Such a logic circuit diagram represents a combinational circuit.

As examples of how to use Boolean algebra to model circuits, we now have a look of two kinds of elementary circuits, and see how the gates introduced above can be used to build them.

**EXAMPLE 7.4.8**    A half-adder is a Boolean circuit that adds two bits, $x$ and $y$, producing two outputs, sum bit $s$ and carry bit $c$, such that $s = 0$ if both $x$ and $y$ have same bit and $s = 1$ otherwise, $c = 1$ if and only $x = y = 1$. In Boolean algebra, $s$ and $c$ can be expressed as $s = (x + y)\overline{(xy)}$ and $c = xy$. The gate diagram for a half-adder is given in Figure 7.4.4.

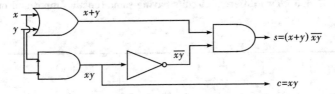

**Figure 7.4.4   Half-adder**

**EXAMPLE 7.4.9**  A **full-adder** is a Boolean circuit that adds three bits ($x, y$ and a carry bit $c$) and produces two outputs (a sum bit $s$ and a carry bit $c'$). The full-adder gate diagram is given in Figure 7.4.5.

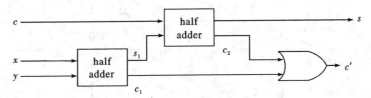

**Figure 7.4.5   Full adder**

Similar to the situation of propositions, different Boolean expressions can have same truth table. In other words a given Boolean function may have different Boolean expressions. Because each Boolean expression can be represented by a circuit diagram, for a given Boolean function there may be many different ways to arrange the gates to get the unique output. The counterpart of this case in circuits is that circuits with different structures can have same function in the sense that same group of input variables will generate same output.

**EXAMPLE 7.4.10**  As functions of $x$ and $y$, the Boolean expressions $xyz + xy\bar{z} + x\bar{y}$ and $y$ have same truth table and therefore yield same output.

**EXAMPLE 7.4.11**  The two circuits in Figure 7.4.6 have same truth table (Table 7.4.3) and therefore they will generate the same output with the same input. Functionally they are the same.

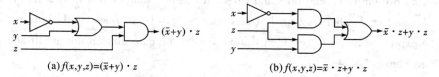

(a) $f(x,y,z)=(\bar{x}+y)\cdot z$  (b) $f(x,y,z)=\bar{x}\cdot z+y\cdot z$

**Figure 7.4.6** Two different circuits having same function input and output

**Table 7.4.3**

| $x$ | $y$ | $z$ | $(\bar{x}+y)z$ | $\bar{x}z+yz$ |
|---|---|---|---|---|
| 0 | 0 | 0 | 0 | 0 |
| 1 | 0 | 0 | 0 | 0 |
| 0 | 1 | 0 | 0 | 0 |
| 1 | 1 | 0 | 0 | 0 |
| 0 | 0 | 1 | 1 | 1 |
| 1 | 0 | 1 | 0 | 0 |
| 0 | 1 | 1 | 1 | 1 |
| 1 | 1 | 1 | 1 | 1 |

In EXAMPLE 7.4.11, although the two Boolean functions perform same functionality, but in (a) only 3 basic gates are used and in (b), 4 basic gates are needed. This differentiation suggests that a Boolean expression can be rearranged in such a way that its implementation only needs least number of gates. This idea is valuable because the reduction in number of gates makes very important sense not only on labor, material and energy saving in digital electronics circuit design and production, but also on improved reliability, and often increased speed. Generally speaking, minimization of circuits is an NP-hard problem, which means it is solvable in polynomial time by a nondeterministic Turing machine. Because any Boolean function can be implemented by a circuit, and conversely any circuit constructed by basic logic gates can be represented by a Boolean function, the task of minimizing circuits is identical to the task of minimizing Boolean functions. We now discuss some basic algorithms for the purpose of minimization of Boolean functions. According to THEOREM 7.4.7, any Boolean function can be written in disjunctive normal Form (DNF). Consequently we have the following concept specifying the meaning of "least number of gates".

**DEFINITION 7.4.10**  A Boolean expression is **minimal** as a DNF if among all equivalent sum-of-products expressions it has the fewest number of summands, and among all sum-of-products expressions with that number of summands it uses the smallest number of literals in the products.

A useful way to obtain the minimal of a Boolean function is to use **Karnaugh map**, also known as a **Veitch diagram**. It was invented in 1952 by Edward W. Veitch, and further developed in 1953 by Maurice Karnaugh. A Karnaugh map for a Boolean expression written in DNF is a diagram that displays the minterms in the Boolean expression. Using Karnaugh maps is a graphical way of using the relationship $AB + A\overline{B} = A$ to simplify a Boolean expression. This is a completely mechanical method of performing the simplification, and so has an advantage over manipulation of expressions using Boolean algebra because it can be performed by computer automatically. Karnaugh maps are effective for expressions of up to about six variables.

The Karnaugh map uses a rectangle divided into rows and columns in such a way that any product term in the expression can be represented as the intersection of a row and a column. The rows and columns are labelled with each term in the expression and its complement. The labels must be arranged so that each horizontal or vertical move changes the state of one and only one variable. The following figure shows the grids for Boolean expressions with two variables ($x$ and $y$), three variables ($x, y$ and $z$) and four variables ($w, x, y$ and $z$).

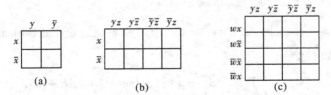

**Figure 7.4.7  Grids for Karnaugh maps**

Each square in each grid above corresponds to exactly one minterm which is the product of the row heading and the column heading. For example, the upper right box in the grid of (a) represents the minterm $x\overline{y}$; the lower right box in the grid of (c) represents

$\overline{w}x\overline{y}z$. As required, the headings are placed in a certain order so that the adjacent squares in any row or column differ in exactly one literal in their headings. The first and last squares in any row or column are to be regarded as adjacent. The variables can be permuted. For example, in (b), the row headings can be $y$ and $\overline{y}$ and the column headings can be $xz, x\overline{z}, \overline{x}\,\overline{z}$, and $\overline{x}z$.

The Karnaugh map for a Boolean expression is obtained by placing a checkmark in each square corresponding to a minterm in the expression.

**EXAMPLE 7.4.12** The Karnaugh maps for Boolean expressions $AB + A\overline{B}$ and $A\overline{B} + \overline{A}B + \overline{A}\,\overline{B}$ are shown in Figure 7.4.8 (a) and (b), respectively.

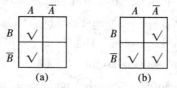

(a)          (b)

**Figure 7.4.8** The Karnaugh map for $AB + A\overline{B}$ and $A\overline{B} + \overline{A}B + \overline{A}\,\overline{B}$

When use Karnaugh map to simplifying a Boolean expression, we firstly create the Karnaugh map. Then draw loops around adjacent pairs of checkmarks. Note that a checkmark may be "in" more than one loop, and a single loop may span multiple rows, multiple columns, or both. For each loop, write an unduplicated list of the terms which appear; For example, no matter how many times $A$ appears, write down only one $A$. If a term and its complement both appear in the list, delete both from the list. Then write the Boolean product of the remaining terms. Finally write the Boolean sum of the products; which is the final simplified expression.

We will now use examples to illustrate the process of simplifying Boolean expressions using Karnaugh maps.

**EXAMPLE 7.4.13** Use Karnaugh map to minimize $A\overline{B} + \overline{A}B + \overline{A}\,\overline{B}$.

*Solution*: Drawing loops on the Karnaugh map of $A\overline{B} + \overline{A}B + \overline{A}\,\overline{B}$ in Figure 7.4.8 (b) results in the following figure.

Because there are two loops, there are two terms in the simplified result. The vertical loop contains $\overline{A}, B, \overline{A}$ and $\overline{B}$. Removing the duplicate $\overline{A}$ produces the list $\{\overline{A}, B, \overline{B}\}$. In the list both $B$ and $\overline{B}$ appear, and so they are cancelled. Then only $\overline{A}$ remains in the list of the vertical loop. From the horizontal loop we remove the duplicate $\overline{B}$, then cancel $A$ and $\overline{A}$ leaving only $\overline{B}$ in the second list. The result is the sum of the remaining terms in two lists, which is $\overline{A} + \overline{B}$. So $A\overline{B} + \overline{A}B + \overline{A}\,\overline{B} = \overline{A} + \overline{B}$.

From the preceding example, we can summarize an algorithm:

## ALGORITHM 7.4.1

1. For a given Boolean expression, write it in the disjunctive normal form.
2. Draw the Karnaugh map for the DNF.
3. Draw loops around adjacent pairs of checkmarks. Checkmarks are "adjacent" horizontally and vertically only, not diagonally. A checkmark may be "in" more than one loop, and a single loop may span multiple rows, multiple columns, or both, so long as the number of checkmarks enclosed is a power of two.
4. For each loop, write an unduplicated list of the terms which appear.
5. If a term and its complement both appear in the list, delete both from the list.
6. For each list, write the Boolean product of the remaining terms.
7. Write the Boolean sum of the products from Step 6; this is the simplified expression.

**EXAMPLE 7.4.14** Simplify $\overline{w}\,\overline{x}y + \overline{w}z\,(\overline{x}y + x\overline{y}) + w\,\overline{x}\,\overline{z} + \overline{w}xy\overline{z} + \overline{w}xy\overline{z}$ using Karnaugh map.

*Solution:* This is an expression in four variables. Using ALGORITHM 7.4.1 we have the following steps:

1. Obtain its DNF is $\overline{w}\,\overline{x}yz + \overline{w}\,\overline{x}y\overline{z} + \overline{w}xyz + \overline{w}\,\overline{x}yz + \overline{w}\,\overline{x}y\,\overline{z} + wxy\overline{z} + w\,\overline{x}\,\overline{z}$
2. Draw its Karnaugh map:

|  | $yz$ | $y\overline{z}$ | $\overline{y}\overline{z}$ | $\overline{y}z$ |
|---|---|---|---|---|
| $wx$ |  |  |  |  |
| $w\overline{x}$ |  | ✓ |  |  |
| $\overline{w}\overline{x}$ | ✓ | ✓ | ✓ | ✓ |
| $\overline{w}x$ | ✓ | ✓ |  |  |

3. Draw loops around adjacent pairs of checkmarks:

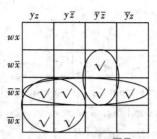

4. The unduplicated lists of the three loops are: $\{\overline{w}\,\overline{x}, yz, y\overline{z}, \overline{y}\,\overline{z}, \overline{y}z\}$, $\{\overline{w}\,x, \overline{w}\,\overline{x}, yz, y\overline{z}\}$ and $\{w\overline{x}, \overline{w}\,\overline{x}, y\overline{z}\}$.

5. The above become to $\{\overline{w}\,\overline{x}\}$, $\{\overline{w}, y\}$, and $\{\overline{x}, y\overline{z}\}$ after cancelling the terms and their complements.

6. The Boolean products of the remaining terms in the three loops are: $\overline{w}\,\overline{x}$, $\overline{w}y$ and $\overline{x}y\overline{z}$.

7. The minimized version is $\overline{w}\,\overline{x} + \overline{w}y + \overline{x}y\overline{z}$.

We conclude the section with a complete example starting from the truth table and ending with the minimized circuit.

**EXAMPLE 7. 4. 15** The following truth table generates a sum-of-products expression with five product terms of three variables each.

| x | y | z | f(x,y,z) |
|---|---|---|----------|
| 0 | 0 | 0 | 0 |
| 0 | 0 | 1 | 0 |
| 0 | 1 | 0 | 1 |
| 0 | 1 | 1 | 1 |
| 1 | 0 | 0 | 0 |
| 1 | 0 | 1 | 1 |
| 1 | 1 | 0 | 1 |
| 1 | 1 | 1 | 1 |

1. write the sum-of-products expression;
2. draw the digital logic circuit of the expression;
3. simplify the expression using ALGORITHM 7. 4. 1;
4. draw the simplified circuit.

*Solution*:

1. $f(x,y,z) = \bar{x}y\bar{z} + \bar{x}yz + xy\bar{z} + xyz + x\bar{y}z$;
2. The circuit diagram is shown in Figure 7.4.9;
3. The Karnaugh map for $f(x,y,z)$ is

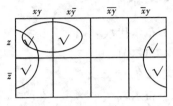

Using ALGORITHM 7.4.1, the simplified result is $y + xz$.

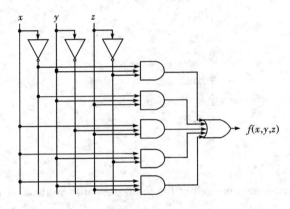

**Figure 7.4.9**　Logic circuit of $f(x,y,z) = \bar{x}y\bar{z} + \bar{x}yz + xy\bar{z} + xyz + x\bar{y}z$

4. The simplified circuit is shown in Figure 7.4.10

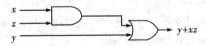

**Figure 7.4.10**　Simplified circuit equivalent to Figure 7.4.4

*Remarks*:

1. Since sometimes there are different ways to draw loops, the minimization of a Boolean expression generally is not unique. For maximum simplification, we make the loops in a Karnaugh map as big as possible. In other words, if we have a choice of making one big loop or two small ones, choose the big loop. The restriction is that the loop must be rectangular and enclose a number of checkmarks that is a power of two.

2. A measure of the complexity of a digital logic circuit is the number of gates and inputs. In EXAMPLE 7.4.15 we have simplified the circuit from nine gates and 23 inputs to two two-input gates. This is a substantial reduction in complexity.

# WORDS AND EXPRESSIONS

| Boolean algebra | 布尔代数 |
| Boolean value | 布尔值 |
| Boolean variable | 布尔变量 |
| Boolean domain | 布尔域 |
| Boolean function | 布尔函数 |
| Minterm | 小项 |
| Maxterm | 大项 |
| Logic gate | 逻辑门 |
| Half-adder | 半加器 |
| Full-adder | 全加器 |
| Karnaugh map | 卡洛图 |

# EXERCISES 7.4

1. Prove that the statements in each of the following pairs are duals of each other:
   a. $a + b = cd, ab = c + d$;
   b. $a + (b + c) = (a + b) + c, a(bc) = (ab)c$;
   c. $a + 1 = 1, a0 = 0$.

2. Verify that the function $f: \{0,1\}^3 \to \{0,1\}$ defined by $f(x,y,z) = x(\bar{z} + \bar{y}z) + \bar{x}$ is a Boolean function in the Boolean variables $x, y, z$. Indicate the minterms and the maxterm in the expression if there are any.

3. Use two different ways to write the Boolean function in preceding exercise in disjunctive normal form.

4. Use two different ways to write the Boolean function in Exercise 2 in conjunctive normal form.

5. Draw a gate diagram for the Boolean function $f$ defined in Exercise 2.

6. Use Karnaugh map to simplify $ABC + A\bar{B}\bar{C} + \bar{A}\bar{B}\bar{C}$. Compare the gate diagrams of the expressions before and after the minimization.

# Reference

1. Kenneth H. Rosen. Discrete Mathematics and its Application[M]. Fifth Edition. 北京:机械工业出版社,2003,9.
2. Ralph P. Grimaldi. Discrete and Combinatorial Mathematics[M]. Third Edition. Addison Wesley Publishing Company,1994.
3. James L. Hein. Discrete mathematics[M]. Jones and Bartlett Publishers,1996.
4. J. A. Bondy, U. S. R. Murty. Graph Theory with Applications[M]. The Macmillan Press Ltd. , 1976.
5. Kenneth H. Rosen. Handbook of discrete and combinatorial mathematics[M]. CRC Press, 2000.
6. Susanna, S. Epp. Discrete mathematics with applications[M]. Third Edition. 北京:高等教育出版社,2005.
7. James A. Anderson. Discrete Mathematics with Combinatorics[M]. 俞正光,陆玫改编. Second Edition. 北京:高等教育出版社,2005.
8. John A. Dossey, Albert D. Otto, Lawrence E. Spence, Charles Vanden Eynden. Discrete Mathematics[M]. 俞正光,陆玫改编. Fourth Edition. 北京:高等教育出版社,2005.
9. 耿素云,屈婉玲,王捍贫. 离散数学[M]. 北京:北京大学出版社,2002.
10. 耿素云,屈婉玲. 离散数学[M]. 北京:高等教育出版社,1998.
11. 左孝凌,李为鉴,刘永才. 离散数学[M]. 上海:上海科学技术文献出版社,1982.
12. 耿素云. 集合论与图论[M]. 北京:北京大学出版社,1997.
13. 李盘林,李丽双,李洋,等. 离散数学[M]. 北京:高等教育出版社,1999.
14. 田丰,马仲梵. 图与网络流理论[M]. 北京:科学出版社,1987.
15. 金晶,徐甫. 离散数学[M]. 北京:科学出版社,2004.
16. 王朝瑞. 图论[M]. 北京:北京工业学院出版社,1987.
17. 王树禾. 图论及其算法[M]. 合肥:中国科学技术大学出版社,1990.